Advances in Solid Oxide Fuel Cells IX

Advances in Solid Oxide Fuel Cells IX

*A Collection of Papers Presented at the
37th International Conference on
Advanced Ceramics and Composites
January 27–February 1, 2013
Daytona Beach, Florida*

Edited by
Narottam P. Bansal
Mihails Kusnezoff

Volume Editors
Soshu Kirihara
Sujanto Widjaja

The
American
Ceramic
Society

WILEY

Cover Design: Wiley

Published by John Wiley & Sons, Inc., Hoboken, New Jersey.
Published simultaneously in Canada.

For general information on our other products and services or for technical support, please contact our
Customer Care Department within the United States at (800) 762-2974, outside the United States at
(317) 572-3993 or fax (317) 572-4002.

Wiley also publishes its books in a variety of electronic formats. Some content that appears in print may
not be available in electronic formats. For more information about Wiley products, visit our web site at
www.wiley.com.

Library of Congress Cataloging-in-Publication Data is available.

ISBN: 978-1-118-80764-4
ISSN: 0196-6219

Printed in the United States of America.

10 9 8 7 6 5 4 3 2 1

Contents

Preface

The tenth international symposium on Solid Oxide Fuel Cells (SOFC): Materials, Science, and Technology was held during the 37th International Conference and Exposition on Advanced Ceramics and Composites in Daytona Beach, FL, January 27 to February 1, 2013. This symposium provided an international forum for scientists, engineers, and technologists to discuss and exchange state-of-the-art ideas, information, and technology on various aspects of solid oxide fuel cells. A total of 85 papers were presented in the form of oral and poster presentations, including twenty invited lectures, indicating strong interest in the scientifically and technologically important field of solid oxide fuel cells. Authors from 22 countries (Austria, Brazil, Bulgaria, Canada, China, Denmark, Egypt, Estonia, France, Germany, India, Italy, Japan, Netherlands, Norway, Portugal, South Korea, Sweden, Switzerland, Taiwan, United Kingdom, and U.S.A.) participated. The speakers represented universities, industries, and government research laboratories.

These proceedings contain contributions on various aspects of solid oxide fuel cells that were discussed at the symposium. Thirteen papers describing the current status of solid oxide fuel cells materials, Science and technology are included in this volume. Each manuscript was peer-reviewed using the American Ceramic Society review process.

The editors wish to extend their gratitude and appreciation to all the authors for their contributions and cooperation, to all the participants and session chairs for their time and efforts, and to all the reviewers for their useful comments and suggestions. Financial support from the American Ceramic Society is gratefully acknowledged. Thanks are due to the staff of the meetings and publications departments of the American Ceramic Society for their invaluable assistance. Advice, help and cooperation of the members of the symposium's international organizing committee (J. S. Chung, Tatsumi Ishihara, Nguyen Minh, Mogens Mogensen, J. Obrien, Prabhakar Singh, Jeffry Stevenson, Toshio Suzuki, and Eric Wachsman) at various stages were instrumental in making this symposium a great success.

We hope that this volume will serve as a valuable reference for the engineers, scientists, researchers and others interested in the materials, science and technology of solid oxide fuel cells.

Narottam P. Bansal
NASA Glenn Research Center

Mihails Kusnezoff
Fraunhofer IKTS

Introduction

This issue of the Ceramic Engineering and Science Proceedings (CESP) is one of nine issues that has been published based on manuscripts submitted and approved for the proceedings of the 37th International Conference on Advanced Ceramics and Composites (ICACC), held January 27–February 1, 2013 in Daytona Beach, Florida. ICACC is the most prominent international meeting in the area of advanced structural, functional, and nanoscopic ceramics, composites, and other emerging ceramic materials and technologies. This prestigious conference has been organized by The American Ceramic Society's (ACerS) Engineering Ceramics Division (ECD) since 1977.

The 37th ICACC hosted more than 1,000 attendees from 40 countries and approximately 800 presentations. The topics ranged from ceramic nanomaterials to structural reliability of ceramic components which demonstrated the linkage between materials science developments at the atomic level and macro level structural applications. Papers addressed material, model, and component development and investigated the interrelations between the processing, properties, and microstructure of ceramic materials.

The conference was organized into the following 19 symposia and sessions:

Symposium 1	Mechanical Behavior and Performance of Ceramics and Composites
Symposium 2	Advanced Ceramic Coatings for Structural, Environmental, and Functional Applications
Symposium 3	10th International Symposium on Solid Oxide Fuel Cells (SOFC): Materials, Science, and Technology
Symposium 4	Armor Ceramics
Symposium 5	Next Generation Bioceramics
Symposium 6	International Symposium on Ceramics for Electric Energy Generation, Storage, and Distribution
Symposium 7	7th International Symposium on Nanostructured Materials and Nanocomposites: Development and Applications

Symposium 8	7th International Symposium on Advanced Processing & Manufacturing Technologies for Structural & Multifunctional Materials and Systems (APMT)
Symposium 9	Porous Ceramics: Novel Developments and Applications
Symposium 10	Virtual Materials (Computational) Design and Ceramic Genome
Symposium 11	Next Generation Technologies for Innovative Surface Coatings
Symposium 12	Materials for Extreme Environments: Ultrahigh Temperature Ceramics (UHTCs) and Nanolaminated Ternary Carbides and Nitrides (MAX Phases)
Symposium 13	Advanced Ceramics and Composites for Sustainable Nuclear Energy and Fusion Energy
Focused Session 1	Geopolymers and Chemically Bonded Ceramics
Focused Session 2	Thermal Management Materials and Technologies
Focused Session 3	Nanomaterials for Sensing Applications
Focused Session 4	Advanced Ceramic Materials and Processing for Photonics and Energy
Special Session	Engineering Ceramics Summit of the Americas
Special Session	2nd Global Young Investigators Forum

The proceedings papers from this conference are published in the below nine issues of the 2013 CESP; Volume 34, Issues 2–10:

- Mechanical Properties and Performance of Engineering Ceramics and Composites VIII, CESP Volume 34, Issue 2 (includes papers from Symposium 1)
- Advanced Ceramic Coatings and Materials for Extreme Environments III, Volume 34, Issue 3 (includes papers from Symposia 2 and 11)
- Advances in Solid Oxide Fuel Cells IX, CESP Volume 34, Issue 4 (includes papers from Symposium 3)
- Advances in Ceramic Armor IX, CESP Volume 34, Issue 5 (includes papers from Symposium 4)
- Advances in Bioceramics and Porous Ceramics VI, CESP Volume 34, Issue 6 (includes papers from Symposia 5 and 9)
- Nanostructured Materials and Nanotechnology VII, CESP Volume 34, Issue 7 (includes papers from Symposium 7 and FS3)
- Advanced Processing and Manufacturing Technologies for Structural and Multi functional Materials VII, CESP Volume 34, Issue 8 (includes papers from Symposium 8)
- Ceramic Materials for Energy Applications III, CESP Volume 34, Issue 9 (includes papers from Symposia 6, 13, and FS4)
- Developments in Strategic Materials and Computational Design IV, CESP Volume 34, Issue 10 (includes papers from Symposium 10 and 12 and from Focused Sessions 1 and 2)

The organization of the Daytona Beach meeting and the publication of these proceedings were possible thanks to the professional staff of ACerS and the tireless dedication of many ECD members. We would especially like to express our sincere thanks to the symposia organizers, session chairs, presenters and conference attendees, for their efforts and enthusiastic participation in the vibrant and cutting-edge conference.

ACerS and the ECD invite you to attend the 38th International Conference on Advanced Ceramics and Composites (http://www.ceramics.org/daytona2014) January 26-31, 2014 in Daytona Beach, Florida.

To purchase additional CESP issues as well as other ceramic publications, visit the ACerS-Wiley Publications home page at www.wiley.com/go/ceramics.

Soshu Kirihara, *Osaka University, Japan*
Sujanto Widjaja, *Corning Incorporated, USA*

Volume Editors
August 2013

DEVELOPMENT OF A PORTABLE PROPANE DRIVEN 300 W SOFC-SYSTEM

Andreas Lindermeir, Ralph-Uwe Dietrich, and Christian Szepanski
Clausthaler Umwelttechnik Institut GmbH (CUTEC)
Clausthal-Zellerfeld, Germany

ABSTRACT

Portable power generation is expected to be an early and attractive market for the commercialization of SOFC-systems. The competition in this market is strong at costs per kilowatt, but weak in terms of electrical efficiency and fuel flexibility. Propane is considered as attractive fuel because of its decentralized availability and easy adaptability to other well-established hydrocarbons, such as camping gas, LPG or natural gas.

The Lower Saxony SOFC Research Cluster was initiated as network project to bundle the local industrial and research activities on SOFC technology. Goal is the development of a stand-alone power supply demonstrator on the basis of currently available SOFC stack technology. Electrolyte supported cells are deployed because to their good stability and robustness. Possible application areas are engine-independent power generation for recreational vehicles or off-grid power supply of cabins and boats. Further potential markets are industrial applications with a continuous demand for reliable power, e.g. traffic management, measuring systems and off-grid sensors and surveillance equipment. To cover the technical requirements of those applications, the SOFC system should provide the following features:

- propane as fuel,
- net system electrical power ≥ 300 We,
- net system efficiency ≥ 35 %,
- compact mass and volume,
- time to full load ≤ 4 hours.

Anode offgas recycle in conjunction with a combined afterburner/reforming-unit in counter flow configuration is used for high efficient fuel gas processing without complex water treatment. All main components are in a planar design and stacked to reduce thermal losses and permit a compact set-up.

INTRODUCTION

The Lower Saxony research cluster consists of 5 institutes from the universities of Braunschweig, Hannover, Clausthal, and the University of Applied Sciences Osnabrück and 9 industrial partners. The project is aimed to introduce innovations in component- and system development by an interdisciplinary team of researchers. The consortium adopted the planar design of the SOFC stack to the other high-temperature components like reformer, stack, heat exchanger and afterburner, thus enabling a compact system setup with a high degree of integration. All units are placed on top or below the stack and rigid pipe connections are avoided whenever possible to minimize space requirements and reduce thermal stresses during the heat-up and cool-down phase. In addition, arrangement and connection of the process units has to consider pressure drop and limitations concerning fabrication and system assembly. The high degree of thermal integration in conjunction with the internal recycle of anode offgas promises an electrical system net efficiency above 35 %. That would be remarkable for a small scale system in the power range < 500 W.

BASIC BLACK BOX CONSIDERATIONS

Prior to the detailed system design and the component specifications, a simple interactive black box model was used to prove the general feasibility of the system concept approach at different boundary conditions (see Figure 1). The Staxera Mk200 stack is

1

equipped with 30 ESC4 cells of H.C. Starck Ceramics and has a rated power output of 700 W_e at a fuel utilization (FU) of 75 %, if operated with a H_2/N_2 mixture of 40/60 Vol.-%[1]. For the basic considerations a performance drop of about 5 % was estimated if syngas from propane reforming was used instead of the H_2/N_2 mixture, resulting in 665 W gross electrical power output. Assuming an electrical gross system efficiency of 60 % the necessary C_3H_8 input is 1,108 W (corresponding 0.022 g/s, HHV). The black box model considers heat losses via the outer system casing by calculating the convective heat flux under the assumption of a flow velocity of the ambient air of v = 0.25 m/s. The heat transfer coefficient α was estimated by

$$\alpha = 2 + 12 \cdot v^{0.5} \tag{1}$$

The heat loss is 300 W for a surface temperature of 50 °C and an area of 1.5 m². The system offgas temperature was calculated by closing the energy balance. The offgas temperature is 121 °C for a cathode air flow rate of 1.2931 g/s (corresponding 60 l_N/min). The cathode air blower causes mainly the parasitic electrical demand of the BoP components. Overall, BoP power demand has to be less than 270 W_e to obtain the demanded electrical net system efficiency of 35 %. For that case, an electrical net power output of 390 W results. Thus, the initial performance goals seem feasible.

Nevertheless, these figures emphasize the impact of heat loss via the surface for small SOFC systems. 27 % of the supplied energy are lost in terms of waste heat at 50 °C surface temperature; 480 W is the convective heat loss at 65 °C surface temperature and the overall energy balance no longer agrees. Additional energy supply would be required to maintain a self-sustaining operation. These simple considerations illustrate the basic necessity of a high degree of thermal integration and the need for internal usage of heat fluxes, what has to be taken into account from the beginning of system design.

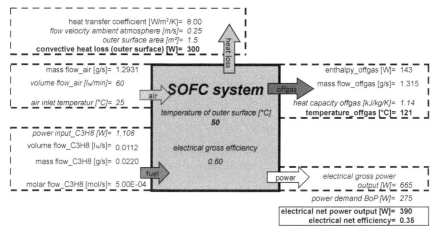

Figure 1. Black box model for proof of feasibility (given input data are formatted italic)

SYSTEM CONCEPT

Figure 2 shows a simplified process flow diagram of the proposed system. Propane and anode offgas (AOG) are fed to the reformer. The AOG contains H_2O, CO_2 and heat from the electrochemical oxidation of the H_2 and CO on the SOFC anode. That is used for endothermic steam- and dry-reforming of the propane. The reformer provides the fuel gas for the SOFC stack. The remaining part of the AOG is fed together with the cathode exhaust air to the afterburner for additional heat generation. Heat is used to maintain the endothermic reformer reactions and for cathode air pre heating to about 650 °C, before entering the stack.

Hot anode offgas recycling is a challenging task and no commercial hardware solution is currently available for the desired flow range. Thus, a piston pump was proposed and developed for AOG recycle. Intercooling of the anode offgas cannot be fully avoided due to the temperature limitations of the compressor bearings and seals. Thus, reheating the compressed AOG with the hot AOG from the anode exit using a tube-in-tube heat exchanger seems to be a reasonable compromise.

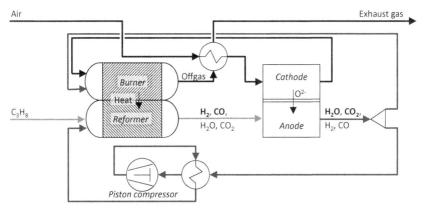

Figure 2. Process flow diagram of the propane SOFC system with anode offgas recycle

The lower limit of the AOG recycle rate is determined by the carbon formation boundary in the reformer for the given temperature. It is shown in previous tests that the oxygen to carbon ratio at the reformer inlet $(O/C)_{Ref}$ (Equation 2) is the key figure with respect to carbon formation[2].

$$\left(\frac{O}{C}\right)_{Ref} = \frac{\dot{n}_{H_2O} + \dot{n}_{CO_2}}{3 \cdot \dot{n}_{C_3H_8}} \tag{2}$$

The parameter $(O/C)_{Ref}$ corresponds to the steam to carbon ratio S/C, well known for steam reforming reactions. Equation 3 and 4 show the strong endothermy of the reforming reactions taken place:

dry-reforming: $\quad C_3H_8 + 3\,CO_2 \rightarrow 6\,CO + 4\,H_2 \qquad \Delta_R^{298}H = 622.1 \;\; kJ/mol \qquad (3)$

steam-reforming: $\;\; C_3H_8 + 3\,H_2O \rightarrow 3\,CO + 7\,H_2 \qquad \Delta_R^{298}H = 498.6 \;\; kJ/mol \qquad (4)$

Figure 3 shows the equilibrium reformate composition at different temperatures based on stationary process flow sheet simulations (ChemCAD®). The flow rate ratio of AOG to propane has been kept constant at a calculated $(O/C)_{Ref}$ of 1.82. Hydrocarbon conversion is almost complete for reforming temperatures above 700 °C. The fraction of H_2 and CO is greater than 60 Vol.-%. Soot formation is inhibited above 720 °C for the distinct operation conditions.

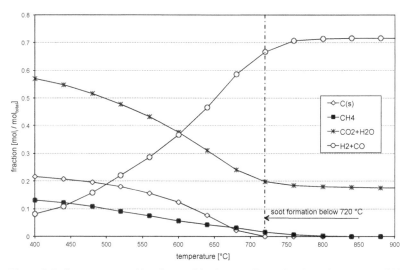

Figure 3. Reformate composition for combined steam-/dry-reforming of propane at different equilibrium temperatures, $(O/C)_{Ref} = 1.82$

While the reformer and burner can be considered as Gibbs reactors (delivering thermodynamic equilibrium values), the flow sheet simulation of the overall process requires the implementation of a confirmed stack characteristic. Key figures for the stack are power output, fuel utilization and electrochemical efficiency at the desired operation point. Thus, a Staxera Mk200/ESC4 stack was evaluated in a stack-test-bench with different fuel gas compositions and throughputs. Figure 4 shows the measured U/I-curves, Table 1 summarizes the stack performance data for the different operation points.

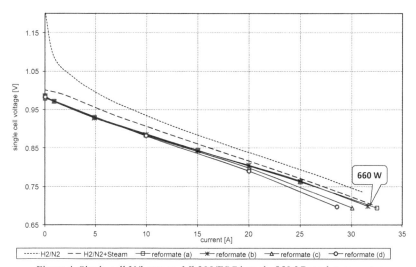

Figure 4. Single cell U/I-curves, Mk200/ESC4 stack, 850 °C stack temperature

Table 1. Experimentally validated stack performance data, Mk200/ESC4 stack, 850 °C stack temperature

Adjustments									Results		
Fuel gas		Composition [Vol.-%]						Flowrate $[l_N/min]$	FU^* [%]	η_e^* [%]	P_e^* $[W_e]$
		H_2	N_2	CO	H_2O	CO_2	CH_4				
H_2/N_2		41.8	58.2	0	0	0	0	23.77	65	40	719
H_2/N_2+steam		38.4	57.8	0	3.8	0	0	24.67	70	41	709
reformate	a	47	0	35	10	7.7	0.3	10.5	71	37	676
	b							**10.34**	**77**	**40**	**660**
	c							9.59	79	41	626
	d							9.06	80	41.5	596

*FU, η_{el} and P are maximum values for each test series

As expected, a fuel gas consisting of H_2 in N_2 gains maximum stack power. A voltage drop was observed throughout the complete current range if the fuel is additionally diluted with steam. Synthetic fuel gas mixtures with a composition derived from thermodynamic simulations of propane reforming with AOG were used to determine the stack performance at the suggested operation point. The stack performance estimations that were used in the black-box model are verified by the experiment. 660 W_e power output was achieved at a fuel utilization of 77 %, resulting in an electrical stack efficiency of $\eta_e = 40$ % (for reformate operation, case b, see Table 1). The expected 5 % drop in stack performance with propane reformat syngas was confirmed.

The system concept was investigated in steady-state operation with ChemCAD^R flow sheet simulations. Figure 5 shows the results in terms of an energy flow chart for one operation point.

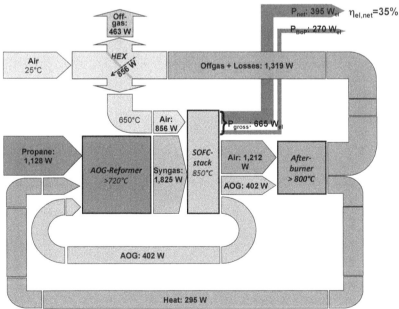

Figure 5: Energy flow chart for the stationary design-point

Propane with an energy content of 1,128 W is fed to the system. Recycled AOG is added, supplying 402 W of heat and chemical energy of the unused syngas components H_2 and CO to the endothermic reformer[3]. Due to the maximum inlet temperature of the AOG-compressor the AOG will be cooled down to 350 °C and successively re-heated as far as possible in real operation. For the sake of simplicity, the respective heat losses are neglected for the energy flow chart. Additional heat flux of 295 W is needed to maintain the steam- and dry-reforming reactions and is supplied by the combustion of the remaining AOG in the afterburner that is in direct thermal contact to the reformer. Thus, chemical energy content of the syngas leaving the reformer is higher than the power of the initial propane feed. The energy content of the syngas of 1,825 W consists of 1,640 W chemical power and 185 W thermal energy. Assuming a stack fuel utilization (FU) of 75 % and an electrochemical efficiency of 54 % the gross power output is 665 W_e. The internal power consumption of the BoP components has to be less than 270 W_e for $\eta_{el} = 35$ % net efficiency, resulting in 395 W_e net system power.

Figure 6 shows a sketch of the assembly concept and a scheme of the gas flows. The main components are enclosed in an inner thermal insulation. Propane enters the system, is mixed with the respective amount of anode offgas delivered by the AOG compressor and fed to the reformer layer. Reformate gas passes the SOFC anode; part of the anode offgas stream is recycled to the reformer inlet, remaining AOG is fed to the catalytic burner that encloses the planar reformer catalyst. The burner exhaust gas enters the heat exchanger for cathode air preheating. Cold cathode inlet air purges the outer casing prior to the entry in the heat exchanger to assure low surface temperatures and thus reduce heat losses through the outer enclosure. The system has only two supply connections, one for the air and propane. Exhaust gas is released by an opening in the outer casing.

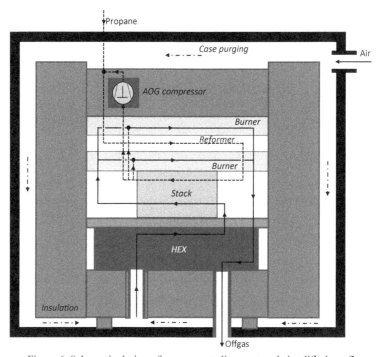

Figure 6. Schematic design of component alignment and simplified gas flows

REFORMER DEVELOPMENT

Supplying sufficient heat for the endothermic reforming reactions is a main task to assure a self-sustaining and soot free reformer operation. As shown in Figure 5, a heat flux of about 300 W has to be transferred from the burner section to the reformer to maintain a reforming temperature above 720 °C. A commercial metallic foil substrate is used as support for the reformer catalyst because of its good heat conductivity, what assures good heat transfer through- and in-plane. A catalyst with an activity for both, steam- and dry-reforming of propane is required. For the proof of proper catalyst choice, a catalytic coated foil package was mounted in a simple housing. For the preliminary catalyst tests, a five-layer substrate was used as support for the catalytic coating. The assembly was equipped with thermocouples (see Figure 7) and placed inside a furnace, heated up to 850 °C and fed with a mixture of propane and a synthetic AOG. The furnace simulates the heat supplied by the anode offgas burner. Reformate gas composition and reformer temperatures are determined for different feed compositions and flow rates. Figure 8 shows the reformer outlet temperature and the methane fraction for different input compositions.

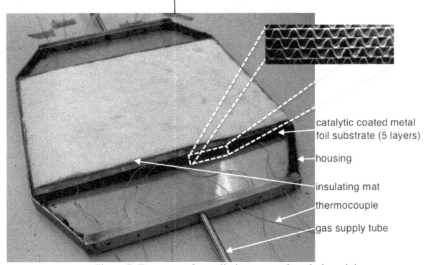

catalytic coated metal
foil substrate (5 layers)

housing

insulating mat

thermocouple

gas supply tube

Figure 7. Test set-up for preliminary test of catalytic activity,
detail: cross-sectional area of the 5-layer metallic support

The methane fraction can be used as indicator for the degree of conversion. The catalyst is well capable of converting propane if sufficient heat is supplied. For propane flow rates above 0.4 l_N/min the reformer temperature falls short of the soot formation boundary. To avoid carbon formation, the tests were terminated at propane flow rates of 0.5 l_N/min.

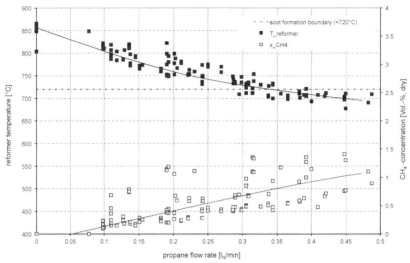

Figure 8: Temperature and CH$_4$ concentration (dry basis) at reformer exit with different C$_3$H$_8$ flow rates, furnace temperature: 850 °C

Good conversion rates at low propane throughput and high reforming temperatures show, that the catalyst is suited for a combined steam- and dry-reforming of propane. For further improved heat transfer through-plane a catalyst substrate with only one layer was chosen for the final reformer set-up and coated with the examined catalyst. The substrate foil has a sheet thickness of 50 μm, resulting in a porosity of 400 cpsi (see Figure 9). The foil package has an overall thickness of 1.41 mm.

Figure 9: Single-layer metallic support for the reformer catalyst

The reformer unit was designed as a joint-part coupling the catalytic afterburner and the reformer layer in a sandwich-design, with the catalytic coated reformer substrate located between two burner plates (Figure 10). A commercial oxidation catalyst was used as burner catalyst, arranged as randomly packed bed above and below the reformer unit.

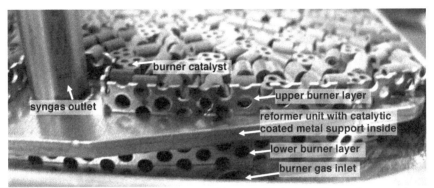

Figure 10: Picture of the reformer-burner-unit in sandwich design with open burner casing

SYSTEM SETUP

 Figure 11 shows a 3D-sketch of the complete system setup. Part of the insulation, the stack compression system and the outer casing are removed for a better view. The AOG-compressor together with the tube-in-tube heat exchanger for the AOG is separated from the hot components by a thermal insulation. The reformer-burner-unit is located on top of the SOFC stack with an adapter plate in between to direct the gas streams in the right manner. Another insulation plate separates the stack from the heat exchanger compartment. The plate heat exchanger is designed as two-stage unit, with a low temperature zone made from alumina and a high temperature unit from Crofer22APU. High temperature steel tubes with bellow compensators connects the burner outlet and the heat exchanger inlet. All other components are connected directly or via the adapter plate.

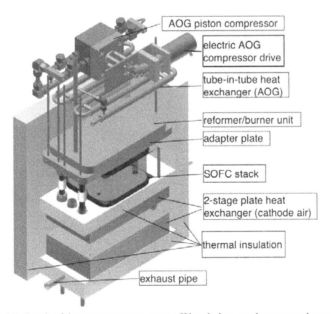

Figure 11: 3D-sketch of the system set-up, parts off insulation, stack compression system and outer casing are removed

CONCLUSION AND OUTLOOK

A concept for a SOFC portable power unit in the power range ≥ 300 $W_{e,net}$ using current stack technology has been presented. A simple black box model proves the general feasibility of the approach and indicates, that an electrical net efficiency $\geq 35\%$ is possible using anode offgas recycling and a high level of thermal integration by planar design of the main components promises. The Mk200 stack is well capable of meeting the system requirements and the planar reformer design with a single-layer metal foil as catalyst support and the chosen catalyst has been proven to work under the ambitious conditions of steam- and dry-reforming of propane with AOG. Next step will be the commissioning of the system and the detailed characterisation at different operation points.

ACKNOWLEDGMENT

The authors would like to thank the colleagues from project partners IAL (University of Hannover), InES (University of Braunschweig), IEE, IMET, ISAF (all University of Clausthal) and LAT (University of Applied Sciences Osnabrück).

This work was financed by "European Regional Development Fund" (ERDF). Financial and advisory support by our industrial partners EcoEnergy GmbH, Solvis GmbH & Co KG, H.C. Starck GmbH, LASER on demand GmbH, SIEB & MEYER AG, GEA AG, EWE AG, Staxera GmbH and Elster Kromschröder GmbH is gratefully acknowledged.

REFERENCES

[1]Staxera GmbH, Product data sheet Article 284, Dresden, Germany, 2008
http://www.staxera.de/fileadmin/downloads/Mk200/081106_PDS_284-Mk200.pdf
[2]S. Chen, C. Schlitzberger, Modeling and Simulation of a Propane SOFC System with Integrated Fuel Reforming Using Recycled Anode Exhaust Gas, Proceedings of the 8th European Solid Oxide Fuel Cell Forum, Luzern, Swiss, 30.6-4.7.2008
[3]R.-U. Dietrich, et al., Using anode-offgas recycling for a propane operated solid oxide fuel cell, Proceedings of the 7th ASME Conference on Fuel Cell Science, Engineering & Technology, Newport Beach, California, USA, 8.-10.6.2009

SOFC-SYSTEM FOR HIGHLY EFFICIENT POWER GENERATION FROM BIOGAS

Andreas Lindermeir, Ralph-Uwe Dietrich, and Jana Oelze
Clausthaler Umwelttechnik Institut GmbH (CUTEC)
Clausthal-Zellerfeld, Germany

ABSTRACT

Power generation from biogas using motor-driven CHP units has a limited electrical efficiency far below 50 %, especially for smaller engines in the power range below 100 kW_e. Fluctuating quality and/or low CH_4 content reduce operation hours and economical and ecological benefit. On the other hand, solid oxide fuel cell (SOFC) systems promise electrical efficiencies above 50 % even for small-scale units and/or low-calorific biogas. Current development tasks of SOFC stack technology are the scale-up of the power range up to the hundreds of kW_e range and further improvements regarding their fuel efficiency, costs and lifetime. Nevertheless commercial state-of-the-art stacks and stack modules are already established in the market and thus available for the evaluation of different system concepts.

In collaboration with the Zentrum für BrennstoffzellenTechnik (Center for Fuel Cell Technology, ZBT) CUTEC has developed a biogas fed 1 kW_e SOFC-system using combined dry and steam reforming of CH_4 for fuel processing. It demonstrates the opportunities of SOFC technology in the biogas market and allows cost-benefit-estimations for future development tasks. To assure a H_2S concentration < 1 ppmv in the reformate gas a sulfur trap was developed based on actual biogas monitoring campaigns. A commercial SOFC stack module with two 30-cell ESC-stacks was used and the system was evaluated with both synthetic biogas mixtures and biogas from the wastewater treatment facility of a sugar refinery. Electrical power output of 850 to 1,000 W_e and electrical gross efficiencies between 39 and 52 % were received for CH_4 contents between 55 and 100 Vol.-%. Results were confirmed during a 500 h test period with synthetic biogas containing 55 Vol.-% CH_4 proving an electric power output of 1,000 W_e and an efficiency of 53 %. No degradation of the stacks or the system components was observed.

INTRODUCTION

Electrical power production from biogas is a constantly growing market especially in Germany with more than 15 TW_eh output in 2010.[1] The installed biogas power capacity rose from 2,291 to 2,780 MW_e during 2011 and the number of installed units increased from 5,905 to 7,100.[2] However, power generation from biogas is still limited by the technical capabilities of conventional CHP plants like:

- Efficiency: The electrical efficiency of a biogas CHP is limited to 30 to 45 %[3] for modern systems and to figures below 35 % for outdated or small-scale (< 100 kW) systems.
- Heat usage: On-site usage of waste heat is required for financial support according to the German feed-in-regulation. Especially for small biogas facilities, this turns out to be challenging and an increase in the power-to-heat ratio would be helpful.
- Biogas quality: A minimum CH_4 content of about 45 Vol.-% is common biogas CHP requirement. Otherwise pilot injection gas engines with higher maintenance costs have to be used. In addition, fluctuations in the biogas quality can cause sudden automatic shutdowns and reduce operation time.

In contrast, high temperature fuel cells like SOFCs provide efficient power generation with power to heat ratios > 1 even at small scale. In addition, they are able to process lean biogases and fluctuating gas composition. If the remaining development tasks will be addressed with sufficient resources, this technology can considerably improve electrical power production from biogas in the near future.

11

To underline the technical potential, a system for power generation from biogas based on a commercial SOFC stack module in the 1 kW$_e$ range was developed and evaluated, both, in the laboratory and at the wastewater treatment facility of a sugar plant. An outstanding power efficiency using biogas with fluctuating composition was demonstrated.

SYSTEM CONCEPT

Downstream of a sulfur trap, a catalytic reformer converts CH$_4$ with CO$_2$, both contained in the biogas, to H$_2$ and CO via dry reforming. Steam is added from an external source to adjust the oxygen to carbon ratio (O/C)$_{Ref}$ required for soot free operation of biogas with fluctuating composition. The (O/C)$_{Ref}$ is defined as

$$\left(\frac{O}{C}\right)_{Ref} = \frac{\dot{n}_{H_2O} + \dot{n}_{CO_2}}{\dot{n}_{CH_4}} \tag{1}$$

This definition is applied because the reformer catalyst supports both the dry reforming (Eq. 2) and steam reforming (Eq. 3) of CH$_4$.

$$CH_4 + CO_2 \rightarrow 2\,CO + 2\,H_2 \quad \text{(dry reforming)} \qquad \Delta_R^{298}H = 247,2 \text{ kJ/mol} \tag{2}$$

$$CH_4 + H_2O \rightarrow CO + 3\,H_2 \quad \text{(steam reforming)} \qquad \Delta_R^{298}H = 206,2 \text{ kJ/mol} \tag{3}$$

Both reactions are highly endothermic so that heat has to be supplied to the reformer catalyst. This is realized by combustion of the anode offgas that still contains sufficient amounts of H$_2$ and CO. The offgas burner is in close thermal contact with the reformer unit and assures a temperature high enough for CH$_4$ reforming without soot formation[4]. Remaining heat of the burner exhaust is used for evaporating and superheating of water.

H$_2$ and CO in the fuel gas are electrochemically oxidized at about 850 °C in a commercial SOFC integrated stack module (Staxera ISM)[5]. Cathode air is heated by an electrical preheater to at least 650 °C and fed to the cathode. Cathode offgas is mixed with burner offgas and exhausted. In a future commercial system a recuperator for air preheating using the cathode offgas heat would be added. Auxiliary components for the air and water treatment and the control unit complete the system. Figure 1 shows a process diagram in which instrumentation and controls are not shown.

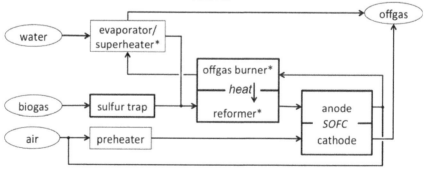

Figure 1. Process flow diagram of the biogas SOFC system (*units developed by project partner ZBT)

STACK CHARACTERIZATION

The ISM contains two Staxera MK200 SOFC stacks with 30 ESC2[6]-cells each and has a rated power output of 1.3 kW_e at the reference point (40 Vol.-% H_2 in N_2). Table 1 summarizes the specifications as given by the manufacturer[5].

Table 1: Specifications of the Staxera ISM with two MK200 stacks[5]

Rated power output	1,300 W[1]
Maximum operation temperature	< 860 °C
Maximum fuel utilization	85%
Dimensions HxWxD	592 x 436 x 331 mm³
Weigth	< 65 kg

[1] at reference point: 40 Vol.-% H_2 in N_2, 75% fuel utilization

A detailed performance map of a single MK200 stack in a defined furnace environment was experimentally determined at the beginning of the project. These data were needed to predict and understand the stack behavior for different biogas compositions and reformer operation regimes and to obtain the essential performance values for the system design.

ZBT provided thermodynamic process simulations to predict the reformate composition after dry and steam reforming of biogas. The simulation parameters CH_4 content in the biogas (55 to 80 Vol.-%) and reformer $(O/C)_{Ref}$-ratio (2.0 to 2.5) were varied. Biogas flow rates are in the range between 1.51 and 2.85 l_N/min[4]. Corresponding fuel gas compositions were mixed at the CUTEC stack-test-bench for the SOFC stack evaluation. Stack core temperature was adjusted to 855 °C by the furnace temperature and the cathode air flow rate. The results of the stack tests are shown in Figure 2. The x-axis is shown in terms of the chemical power of the biogas fed to the system, based on the LHV of CH_4.

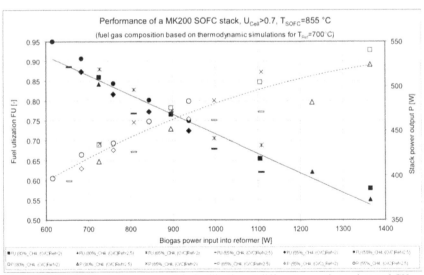

Figure 2. Maximum stack power and fuel utilization at different operating points and CH_4 content of the biogas. Fuel gas composition was adjusted on the basis of thermodynamic simulations with a reformer outlet temperature of 700 °C[4]

The MK200 stack yields 540 W_e at 58 % fuel utilization from reformate gas generated from 2.85 l_N/min biogas with 80 Vol.-% CH_4 and $(O/C)_{Ref}$ = 2, corresponding to a biogas power input of 1,364 W. On the contrary, fuel utilization of more than 90 % was reached at a lower power output of 420 W_e for a weaker biogas (e.g. 55 Vol.-% CH_4) at the same biogas flow rate. Increasing the $(O/C)_{Ref}$ ratio in the reformer shifts the stack performance only for the lean biogas towards higher values of fuel utilization and stack power.

For higher CH_4 content, more steam and CO_2 at the anode reduce power and fuel utilization of the stack, thus a small $(O/C)_{Ref}$ ratio should be preferred for high system efficiency. On the other hand, the risk of carbon formation in the reformer increases with lower $(O/C)_{Ref}$, so a compromise between system efficiency and a safe operation has to be found. Thermodynamic calculation indicated that for soot free operation an $(O/C)_{Ref}$ above 2.5 at temperature > 625 °C has to be assured[4].

Stack operation at maximum power is not recommended for high system efficiency due to the reduced fuel utilization. At lower biogas flow rates SOFC fuel utilization and hence system efficiency increase, but bigger stacks with higher investment costs are required for the desired power output. The economic optimum between investment costs and operating profit has to be found for each application. However, it is suggested to keep the fuel utilization of the stack above 70 % to demonstrate the efficiency gain compared to common CHP-units. Here, the system design point was set to 5.14 l_N/min biogas (assuming 65 Vol.-% CH_4), corresponding to a biogas energy input of 1,998 W (based on LHV of the biogas). Stationary system simulations from ZBT and the results from the stack characterization predict a system power output of 924 W_e. At a slightly reduced fuel utilization of 68 % the electrical stack efficiency will be 37.4 %, and the corresponding system efficiency 46.2 %. Because SOFC stack technology still lacks sufficient lifetime for commercial applications a stationary stack test at the following conditions was performed to evaluate the degradation:

- fuel gas flow rate: 9.97 l_N/min,
- fuel gas composition according to thermodynamic equilibrium (x_{CH4} = 65 Vol.-%, T_{Ref} = 700 °C and $(O/C)_{Ref}$ = 2.5),
- constant stack current of 20 A (corresponds to FU = 56 %),
- stack core temperature: 855 °C.

The results of the long-term test are shown in Figure 3. No degradation could be detected within 1,300 hours, even with a few load dumps caused by equipment failures.

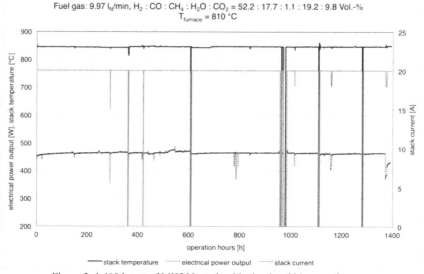

Figure 3. 1,400 h test of MK200 stack with simulated biogas reformate

BIOGAS MONITORING AND SULFUR REMOVAL

An online gas analysis system was installed at the biogas plant of the Nordzucker sugar refinery wastewater treatment in Uelzen, Germany, to determine the concentration range and changes of the relevant biogas components CH_4, CO_2 and H_2S. Sugar plants operate in yearly campaigns typically lasting from September to January. Stable biogas production starts with a delay of some weeks needed for the stabilization of the biological processes.

Online monitoring of the biogas composition was conducted during the sugar campaigns 2009/10, 2010/11 and 2011/12. Additionally, biogas samples were analyzed offline by GC-MS-, IC- and ISE-analysis to detect possible contaminants harmful to the reformer catalyst and/or the SOFC. As result, no siloxane, BTEX aromatics (detection limit of 0.002 mg/m^3), chlorine (detection limit: 2 mg/m^3) and fluorine (detection limit: 0.6 mg/m^3) were detected. Only H_2S needs to be removed from the biogas. Results of the biogas monitoring with the online analysis for the three campaigns are shown in Table 2.

Table 2: Results of the biogas online monitoring
during Nordzucker sugar campaigns

Campaign 2009/2010	Gas concentration range (dry basis)
CH_4	62 – 73 Vol.-%
CO_2	28 – 36 Vol.-%
H_2S	4 – 53 ppmv
2010/2011	
CH_4	58 – 73 Vol.-%
CO_2	23 – 41 Vol.-%
H_2S	13 – 475[1] ppmv
2011/2012	
CH_4	59 – 72 Vol.-%
C	29 – 37 Vol.-%
H_2S	40 – 2.400[2] ppmv

[1] at the campaign start, [2] at the campaign end

The CH_4 and CO_2 concentration ranges are comparable for the investigated campaigns showing mean values of approx. 67 Vol.-% and 33 Vol.-% for CH_4 and CO_2 respectively. On the contrary, the H_2S content depends strongly on the operation regime of the biogas plant. Especially at the beginning and the end of the campaign, the H_2S concentration exceeds those under steady-state operation by orders of magnitude. However, within the stable phases the H_2S content was below 100 ppmv. Figure 4 shows the measured biogas composition during nearly 4 weeks of the sugar campaign 2010/11.

The SOFC system requires desulfurization of the biogas to below 1 ppmv to avoid poisoning of the reformer catalyst and the SOFC anode. A sulfur trap was designed and evaluated using a commercial H_2S adsorbent based on sulfide forming agents (extrudate, $1 = 5...10$ mm, d = 1.6 mm). The following manufacturer data for the adsorbent are available:

- gas hourly space velocity range: GHSV = 500...1,500 h^{-1}
- length to diameter ratio for packed bed: $1 / d \geq 3$
- load capacity for H_2S at 1 ppmv breakthrough: c = 15.8 % w/w

Breakthrough curves were recorded in the laboratory and used to design a sulfur trap. This sulfur trap was tested at the Nordzucker biogas plant during the campaign 2010/11. No H_2S breakthrough was detected within 22 days of operation, though the inlet H_2S

concentration was higher than 450 ppmv at the beginning of the test. The calculated sulfur load of the adsorbent after the test campaign was 13 % w/w[4].

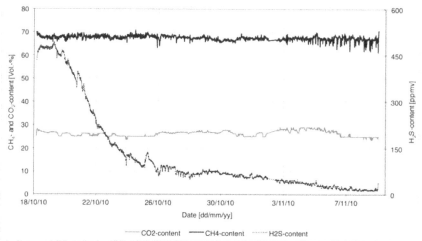

Figure 4: Results of the online monitoring of biogas during sugar campaign 2010/11

OPERATION OF THE SOFC-SYSTEM

The system units were assembled together with the necessary instrumentation and BoP components and put into operation. Figure 5 shows a picture of the system with several insulations removed for better view.

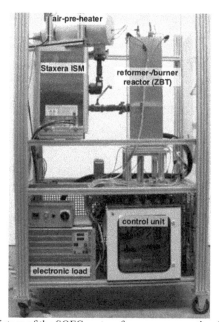

Figure 5: Picture of the SOFC system for power generation from biogas

System tests were performed operating pure steam reforming of 100 Vol.-% CH_4. Subsequently, different biogas compositions were fed by adding CO_2 to the system. Performance data are shown in Figure 6.

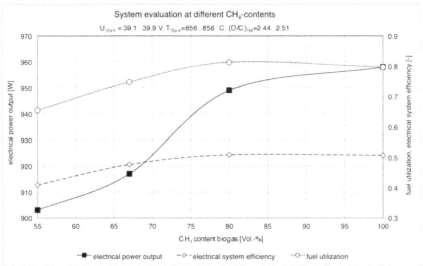

Figure 6: Performance results for laboratory system tests with different CH_4 contents

Heat production by the offgas burner and the heat usage of the endothermic reformer unit were balanced for all operation points to ensure thermal self-sustaining operation. The electrical power output increases with higher CH_4 content in the fuel gas mixture. Fuel utilization rises up to a maximum of about 80 % for high CH_4 content. Higher CH_4 content reduces the biogas flow required for thermal balancing, thus decreasing the amount of reformate gas. Accordingly, fuel utilization improves and hence overall electrical efficiency. Electrical power of 900 to 960 W_e and a system efficiency ranging from 41 to 51 % were received.

These figures indicate that the system is capable of converting biogas with high electrical efficiency over a broad range of biogas composition. Table 3 compares values for the predicted operation point with results of the experimental evaluation.

Table 3: Comparison between predicted and experimentally validated system results

Parameter	Adjusted/ Predicted	Experimentally validated
CH_4 content biogas	65 Vol.-%	67 Vol.-%
$(O/C)_{Ref}$	2.5	2.51
Biogas flow rate	5.14 l_N/min	4.8 l_N/min
Chemical power input	1,998 W	1,924 W
Fuel utilization SOFC	0.68	0.75
Electrical system power output	924 W_e	917 W_e
Gross electrical system efficiency	46.3 %	47.7 %

The experimental system efficiency is higher than predicted, mainly due to the improved fuel utilization of the stack at the specific operation point. System power output matches the predicted value though biogas flow rate and thus energy input was lower.

Following the lab characterization, the system was operated with biogas from the Nordzucker wastewater treatment facility containing approx. 68 Vol.-% CH$_4$. Figure 7 shows the electrical power output and the resulting system gross efficiency for different flow rates.

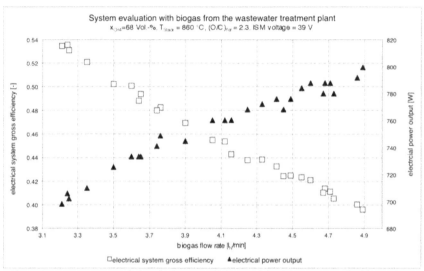

Figure 7: Power output and electrical gross efficiency of the SOFC system for biogas from the Nordzucker wastewater treatment plant

700 to 800 W$_e$ were generated at a system efficiency of 40 to 54 %. Compared with lab conditions, power decreases from approx. 920 W to 780 W$_e$. The system was operated thermally stable for different input parameters at the given fluctuating biogas composition. All relevant parameters of H$_2$S content, temperature and pressure were within the demanded range and the whole system was in a thermal self-sustaining mode at the biogas plant (except the electrical air preheating of cathode air).

Because of the distinct power loss between the lab test and the on-site experiments, the SOFC ISM was replaced by a new one. The stack manufacturer investigated reasons for the power loss and suggested the high fuel utilizations during the start up of the system at the biogas plant as main reason for the degradation.

Additional measurements with the new ISM were conducted with a synthetic biogas in the lab. Here, a 55/45 Vol.-% mixture of CH$_4$ and CO$_2$ was used in contrast to the high CH$_4$ content of the Nordzucker biogas. Such composition is more representative for biogas based on renewable primary products like corn silage. The system was operated for nearly 500 h with results shown in Figure 8.

Power output of 1,000 W$_e$ and electrical system gross efficiency of 52 % were obtained after the system had thermally stabilized. The system was switched to hydrogen operation for two short time periods after approx. 160 and 230 h because of valve malfunctions. In spite of these events no decrease in power output or system efficiency was detected within the test period. The performance values of the initial laboratory system tests were confirmed and even exceeded with the new ISM.

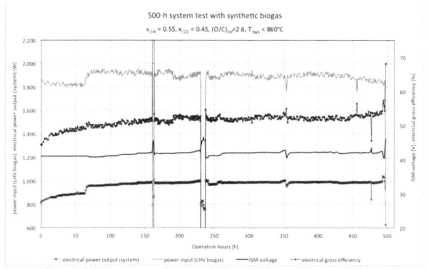

Figure 8: SOFC-system performance during a 500 h test with synthetic biogas

DISCUSSION AND ECONOMIC CONSIDERATIONS

Combined dry and steam reforming of biogas in conjunction with commercial SOFC stack technology has been proven as a valid concept for highly efficient power generation from biogas. The 500 h test result with synthetic biogas with an electrical system gross efficiency of 52 % can be taken as the current technology status in the 1 kW$_e$ power range. Heat flux for the endothermic reforming reaction is provided by combustion of the offgas, thus avoiding additional biogas consumption.

With the technical validation of the concept one requirement of the market introduction of biogas fed SOFC-systems is fulfilled. Remaining tasks are:

- scale-up of SOFC stack technology to match biogas plant size,
- reduction of specific system costs per kW$_e$ by increasing stack module power,
- build-up of system production lines,
- demonstration of long-term system lifetime and maturity.

According to a report from Pike Research German biogas equipment manufacturers and project developers are expected to export their expertise abroad to tap opportunities across Europe, the Americas, and Asia Pacific. Global installed production capacity is now more than 800 billion cubic feet per year (22.6 billion m^3), representing nearly 14.5 GW$_e$ of installed power capacity. This market reached $17.3 billion in global revenue in 2011 and will nearly double to $33.1 billion by 2022[7].

The 7,215 biogas plants installed in Germany have an average power of about 400 kW$_e$.[2] 1,300 biogas plants newly connected to the grid in 2011 have an average power of about 420 kW$_e$ showing a trend to ever bigger power stations.[2] SOFC technology is still behind that unit capacity trend with the current maximum of 210 kW$_e$ base load output (net AC) of the Bloom Energy ES-5700 Energy Server[8]. The Bloom Energy Server uses directed biogas, so intrinsic CO_2 has to be removed rather than being used as oxidant as in the concept presented here.

Beside the global trend the specific plant size depends on the local availability of biogas and on political preferences and support. In Germany the recent revision of the Renewable Energy Sources Act (EEG2012) offers a special feed-in tariff for small-scale biogas plants

< 75 kW$_e$. If that market will establish, far more units could be sold worldwide and other SOFC system developers might serve that upcoming market.

If current CHP technology for biogas is compared to the SOFC concept presented here the allowable premium price of the latter can be estimated by calculating the benefits of higher power sales. Assuming a 75 kW$_e$ SOFC system running for 20 years with an electrical efficiency of 52 %, power can be sold at a price of 0.25 €/kWh$_e$ according the EEG2012 feed-in tariff. Table 4 shows the basic input data for the economical comparison from literature if available or reasonable estimates otherwise.

The price target in terms of specific investment costs for a biogas SOFC system can be estimated by comparison of the economic benefit compared to standard CHP units.

Table 4: Basic input data for the economical comparison of conventional CHP and SOFC

	Conventional CHP	SOFC-system
Available biogas	26.26 Nm^3/h	
CH$_4$ content biogas	55 %	
Biogas input	144 kW	
Proceeds from process heat	0.02 €/kWh	
Feed-in tariff	0.25 €/kWh [3]	
Biogas production costs	0.05 €/kWh	
Operation hours (full load)	7,000 h/a	
Electrical efficiency	36 % [9]	52 %
Electrical power	52 kW	75 kW
Overall efficiency	85 %	
Thermal power	70.6 kW	47.5 kW
Specific investment costs	2,039 €/kW [9]	Target figure (tbd)
Specific maintenance costs	0.0275 €/kWh [9]	
Operation time to general overhaul	every 40,000 hours (full load)	
Specific costs for general overhaul	300 €/kW [10]	50 % of investment cost (for stack replacement)[11]

Using these numbers an upper limit of 5,000 €/kW$_e$ (6,300 $/kW$_e$) is estimated for the target figure. Thus the specific SOFC investment costs limit is 2.5 times higher as the current price level for a standard CHP unit. Since current development goals for stationary SOFC applications are well within this range[11] further steps towards the introduction of SOFC-systems into the biogas market are recommended. Open issues are scale-up of stack technology and demonstration of viable system lifetime.

SUMMARY AND OUTLOOK

An SOFC-system for power generation from biogas was developed and characterized demonstrating the efficiency potential of the combined dry and steam reforming concept in conjunction with state-of-the-art SOFC technology. The stack characteristic and the performance of the reformer module were determined with respect to the biogas composition. The complete system was dimensioned using the results of the component testing and the stationary process simulations. The overall system was mounted as a stand-alone unit and evaluated at different operation points under lab conditions. Subsequently it was operated successfully on site with actual biogas. Since the performance values obtained before under lab conditions could not be reached the stack module was replaced by a new one of the same type. After this, under lab conditions power output and system efficiency slightly exceeded the earlier results and was operated for more than 500 h without any observable degradation. The following main results were obtained:

- The biogas-SOFC-system works stable over a broad range of CH_4 content.
- Fluctuating concentrations can be controlled by addition of sufficient steam to the reforming unit.
- Power output and system efficiency projections are proven in a long-term test under lab conditions.
- Promising performance values of $1 kW_e$ and 52 % efficiency for a simulated biogas containing 55 Vol.-% CH_4 are shown.

ACKNOWLEDGMENT

The authors would like to thank their colleagues from project partner ZBT for their work concerning the stationary flow sheet simulations and the development of the reformer and afterburner unit. Additional thanks are given to Staxera GmbH and Nordzucker AG for their technical support during the project.

This work was financed with funds of the German "Federal Ministry of Economics and Technology" (BMWi) by the "Federation of Industrial Research Associations" (AiF) within the program of "Industrial collective research" (IGF-project no. 16126 N). Funding received from "DECHEMA Gesellschaft für Chemische Technik und Biotechnologie e.V." (Society for Chemical Engineering and Biotechnology) is also gratefully acknowledged.

REFERENCES
[1]D. Böhme, W. Dürrschmidt, M. van Mark (Red.): Erneuerbare Energien in Zahlen – Nationale und Internationale Entwicklung, Bundesministerium für Umwelt, Naturschutz und Reaktorsicherheit (BMU), June 2010
[2]Fachverband Biogas e.V., http://www.biogas.org/edcom/webfvb.nsf/id/DE_Branchenzahlen/$file/11-11-15_Biogas%20Branchenzahlen%202011.pdf, November 2011)
[3]Fachagentur Nachwachsende Rohstoffe e.V. (FNR): Faustzahlen Biogas, http://www.bio-energie.de/biogas/faustzahlen_1/, November 2010
[4]R.-U. Dietrich, A. Lindermeir, J. Oelze, C. Spieker, C. Spitta, M. Steffen: SOFC power generation from biogas – improved system efficiency with combined dry and steam reforming, 219th ECS Meeting, Montreal (Canada), May 1 - 6, 2011
[5]Sunfire GmbH: Product website Integrated Stack Module (ISM), http://www.sunfire.de/?page_id=135, April 2012
[6]H.C. Starck Ceramics GmbH, Product brochure; http://www.hcstarck.com/hcs-admin/file/d329bce42e52fd5b012fbf2a100a1d06.de.0/ESC2.pdf
[7]Pike Research's report "Renewable Biogas"; http://www.pikeresearch.com/research/renewable-biogas
[8]Bloom Energy: ES-5700 Energy Server datasheet; http://www.bloomenergy.com/fuelcell/es-5700-data-sheet/
[9]Fachagentur Nachwachsende Rohstoffe e. V. (FNR): Biogas; http://mediathek.fnr.de/media/downloadable/files/samples/b/i/biogas_web.pdf, 2012
[10]ASUE - Arbeitsgemeinschaft für sparsamen und umweltfreundlichen Energieverbrauch e.V.: BHKW-Kenndaten 2011 – Module, Anbieter, Kosten, Stand: Juli 2011, http://asue.de/cms/upload/broschueren/2011/bhkw-kenndaten/asue-bhkw-kenndaten-0311.pdf
[11]C. Wunderlich: Wertschöpfung in der Brennstoffzellenindustrie auf Basis sächsischen F&E know-hows, 2. Dresdner Wasserstofftag, Dresden, 9. Mai 2006, http://www.uzdresden.de/fileadmin/user_upload/downloads/Brennstoffzellen_in_Sachsen_Wunderlich.pdf
[12]M. Hauth, J. Karl: Entwicklungsstand innovativer SOFC Systeme zur dezentralen Energiebereitstellung, 11. Symposium Energieinnovation, 10.-12.02.2010, Graz, Austria

Development of Solid Oxide Fuel Cell Stack Modules for High Efficiency Power Generation

Hossein Ghezel-Ayagh
FuelCell Energy, Inc.
3 Great Pasture Road, Danbury
Connecticut, 06813, USA

ABSTRACT
    FuelCell Energy is developing SOFC systems and technology for very efficient, economically viable, coal-based power plants through a cooperative agreement with the U.S. Department of Energy. A key project objective is implementation of an innovative system concept in design of a Baseline Power Plant (>100 MW) to achieve an electrical efficiency of >55% based on coal higher heating value, exclusive of power requirements for CO_2 compression. The Baseline system is designed to remove at least 90% of carbon in the syngas for sequestration as CO_2. The coal-based power plant is targeted to have a cost of $700/kW (2007 US dollars) for the SOFC power block.

INTRODUCTION
    Integrated Gasification Fuel Cell (IGFC) power plants, incorporating solid oxide fuel cell (SOFC), are an attractive alternative to traditional pulverized coal-fired steam cycle (PC) and Integrated Gasification Combined Cycle (IGCC) power plants, for central power generation. Key features of the IGFC power plants are: high efficiency, near-zero emissions of SOx and NOx, low water consumption, and amenability to carbon dioxide (CO_2) capture. Recent SOFC technology advancements and increased power density are indicative of the cost-competitiveness of IGFC systems relative to other power generation technologies utilizing coal.
    FuelCell Energy (FCE) is developing SOFC systems and technology for very efficient, economically viable, coal-to-electricity power plants (utilizing synthesis gas from a coal gasifier) through a cooperative agreement with the U.S. Department of Energy (DOE) Office of Fossil Energy's Solid State Energy Conversion Alliance program (1, 2, 3). One of the key objectives of the multiphase project is implementation of an innovative system concept in design of a Baseline Power Plant (>100 MW) to achieve an electrical efficiency of >55% based on higher heating value (HHV) of coal, exclusive of power requirements for CO_2 compression. The Baseline system is designed to remove at least 90% of carbon in the synthesis gas (syngas) for sequestration as CO_2. The coal-based power plant is targeted to have a cost of $700/kW (2007 US dollars) for the SOFC power block.
    FCE utilizes the planar SOFC technology of its (wholly owned) subsidiary, Versa Power Systems, Inc. (VPS), for all its SOFC development programs. SOFC technology development includes cell performance enhancement, stack building block scale-up, stack robustness improvement, stack tower design and testing, and development of a 60 kW module accommodating multiple stacks. FCE is utilizing its high temperature fuel cell design and operational experience in the development of scaled up SOFC stack towers and stack modules based on the building block units manufactured by VPS.
    In addition to the 500 MW (nominal) Baseline Power Plant system, a small-scale (60 kW nominal) Proof-of-Concept Module (PCM) system is also being developed. The 60 kW natural gas fueled system is suitable for dispersed power generation applications. FCE efforts (currently in Phase III of the project) are focused on the stack tower and module development, and the development of the PCM system.

INTEGRATED GASIFICATION FUEL CELL POWER PLANT SYSTEM DEVELOPMENT

Advanced Baseline IGFC Power Plant system has been developed. Figure 1 shows a simplified block flow diagram of a typical system. The system employs catalytic gasification and warm gas cleanup to provide syngas fuel for the SOFC (primary power generator for the plant). The system also employs oxycombustion of the anode exhaust (SOFC) for CO_2 capture. The post-fuel cell capture of CO_2 derives benefits from CO shift occurring in the SOFC. Though not shown in Figure 1, a syngas expander and an anode recycle are included in the plant. The syngas expander is used to generate supplemental electric power as the clean syngas is expanded to near-atmospheric pressure prior to the fuel cell feed. The anode recycle is utilized to enable system operation at high overall fuel utilization and recycle some of the water produced in the fuel cell. The SOFC system consumes 75% less water compared to PC plants (using scrubbing technology for carbon capture). The plant is well integrated to achieve high electrical efficiency. The heat recovery steam generator provides the steam for the bottoming cycle which also generates supplemental electric power. The system offers an electrical efficiency of 58.7% based on HHV of coal, while capturing > 99% of carbon (in the syngas) as CO_2.

The conceptual layout of the power island was prepared to facilitate the Factory Cost estimation. As shown in Figure 2, the power island consists of 8 sections each containing 42 fuel cell modules. The module Sections are grouped with 4 Sections on each side of the centralized Power Island equipment. The footprint of the IGFC plant was found to be comparable to an IGCC plant. Recent Factory Cost Estimates have shown an SOFC power island cost of $635/kW in 2007 US dollars, meeting the DOE cost target of ≤ $700/kW.

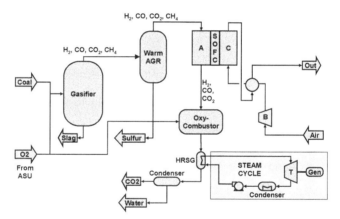

Figure 1. Block Flow Diagram of Baseline IGFC Power Plant

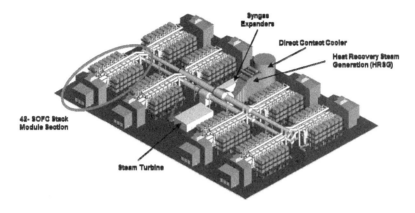

Figure 2. Baseline IGFC Plant Power Island Layout

PROOF-OF-CONCEPT MODULE SYSTEM DEVELOPMENT

A pipeline natural gas (NG) fueled SOFC PCM system is being developed. The design activities are focused on mechanical and electrical balance-of-plant (BOP) systems that will serve as a platform for PCM tests. This plant will have a 60 kW SOFC module and the BOP supporting the fuel cell module. The PCM system will be a stand-alone, modular (skid-mounted, shippable), outdoor-rated design. Figure 3 shows a simplified block flow diagram of the PCM system. PCM system includes anode recycle which provides steam needed for reforming of natural gas fuel, thereby eliminating the need for an external water supply during the plant operation.

A system configuration improvement has been implemented in the design process to enhance combined heat and power (CHP) capabilities of the 60 kW PCM system. High temperature heat available in the SOFC plant exhaust stream is an important attribute when considering the overall system efficiency and plant economics. The key improvement to the system is the reconfiguration of the cathode air flow path for the SOFC stacks located within the module resulting in a higher temperature exhaust stream (~470°C) which is suitable for steam generation. Process modeling and simulations of the reconfigured PCM system have been performed for the full power, rated power, and heat-up modes of operation. Table I summarizes performance for the full power case. The electrical efficiency for the full power case is estimated to be 61.8% based on lower heating value (LHV) of natural gas. The reconfigured PCM system has the potential to offer nearly 89% overall (thermal plus electrical) efficiency for CHP applications.

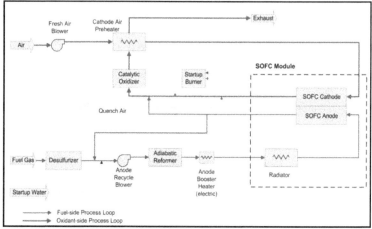

Figure 3. 60 kW PCM System Block Flow Diagram

Table 1. 60 kW PCM System Performance Estimate at Full Power Conditions

PCM System Performance Summary						
POWER GENERATION PROFILE:				**SOFC EFFICIENCY (Fuel to Electricity)**		
Fuel Cell DC Power	**70.2**	**kW**		Electrical Efficiency (LHV)	61.8	%
Inverter Efficiency	95.0	%		Electrical Efficiency (HHV)	55.7	%
Inverter Loss	3.5	kW				
SOFC Gross AC Power	66.7	kW		**CHP**		
Prime Movers	1.5	kW		**Available Heat To 250°F**	77546	BTU/hr
Parasitic AC Power	1.3	kW			22.7	kW
Net AC Power Consumption	2.8	kW		**Available Heat To 120°F**	95715	BTU/hr
Net SOFC Plant AC Output	63.9	kW			28.1	kW
Energy Input				**CHP EFFICIENCY**		
Fuel Flow	17.31	lb/hr		Total Efficiency (LHV) to 250°F	83.8	%
Fuel Low Heating Value	20392	BTU/lb		Total Efficiency (HHV) to 250°F	75.5	%
Fuel Energy (LHV)	103.4	kW				
Fuel High Heating Value	22610	BTU/lb		Total Efficiency (LHV) to 120°F	88.9	%
Fuel Energy (HHV)	114.6	kW		Total Efficiency (HHV) to 120°F	80.2	%

SOFC TECHNOLOGY DEVELOPMENT AND MODULE DESIGN

In parallel to FCE's system development and SOFC module design activities, VPS has also made significant progress in the cell and stack development areas. Toward improving the cell and stack robustness, a hydrothermally stable thin anode substrate cell was developed (which also offers a substantial cost reduction). The design was validated in a 16-cell (550 cm2 cell active area) stack test by completing five thermal cycles while progressively increasing anode gas humidity (to the most stringent conditions) during heat up and cool down. The standard cell manufacturing process was modified based on the hydrothermally stable anode substrate material developments. The cell manufacturing process yield of >90% has been regained based on fabrication of over 1000 cells (25 cm x 25 cm size).

SOFC stack development focused on increasing stack reliability and endurance. Stack tests were conducted to further define and design a standardized stack building block unit. Multiple 96-cell (550 cm2 cell active area, 15 kW nominal) stacks were fabricated, successfully factory conditioned and acceptance tested. One stack was tested for >5000 h with an average performance degradation rate of <1.6%/1000 h. The results were almost duplicated in a second stack which showed a degradation rate of <1.4%/1000 h over 3500 h of operation.

The stack tower concept suitable for large-scale SOFC modules has been successfully demonstrated. A test was conducted on a stack tower built using two 92-cell blocks (previous block size). The tower was tested in a module enclosure environment, with fuel compositions representative of the system (simulated Baseline Power Plant fuel gas). The peak power test achieved 30.2 kW DC output. Following the peak power test, during steady state endurance testing under the Normal Operating Conditions, the stack tower produced an output of 27.0 kW. Overall, the stack tower operated for over 1900 hours. Subsequently, a 30 kW stack tower was assembled using two 96-cell stack blocks. Figure 4 shows a picture of the completed stack tower. The tower test included performance characterization using a simulated natural gas-based fuel feed. A parametric study involving operation around 68% fuel utilization (per pass) was conducted. Endurance testing at 62% fuel utilization was then initiated.

Figure 4. 30 kW SOFC Stack Tower : Assembled Using Two 96-Cell Stack Building Blocks

The SOFC module development focused on the design of a 60 kW module, based on an array of four stacks (quad), each stack containing a 96-cell block. The stacks in the quad are assembled on a single fixed-end base. The quad base design provides the support structure and facilitates the gas flow distribution to the stacks. The module design consists of highly engineered components that require verification for functionality and performance. To test the hardware components as assembled, a (60 kW) module utilizing mock (non-working) stacks was constructed. The major components included the quad base, high temperature flanges, fuel gas (radiative) heat exchanger (HX), conductive module gaskets, hot buss bars, stack compression springs and vessel/module enclosure. The mock stack quad assembly was subjected to cold flow tests for measurements of gas flow uniformity among the stacks. Subsequently, the assembly underwent testing in hot environment similar to the actual system conditions. The hot test results showed that the current design provides very good flow distribution. Performance of the electrical (configuration) system for the module, including the dielectric (isolation) component performance, was also verified. Another important accomplishment of the module hot test was monitoring/characterization of the anode gas superheater (radiative) performance. Based on the hot test results, the performance projection for the HX with actual operating stacks was verified to exceed the design target. The tests showed that the module and the non-repeat component designs (mechanical and electrical) meet the system requirements. The operability and robustness of the 60 kW stack module design were verified.

After successful verification of module components in the mock stack quad assembly module test above, a 60 kW module accommodating four real SOFC stacks was assembled and testing was initiated. Figure 5 shows the module with and without enclosure. The left-side photograph was taken during module assembly before installation of the module enclosure. The right-side photograph shows the module with enclosure, installed in the power plant test facility and undergoing testing. Cathode exhaust temperatures of the individual stacks during the module heat up were very uniform indicating the uniformity of flows and heat loss within the module enclosure. Individual stack open circuit voltages at the end of heat up further confirmed this. Grid-connected (on-load) operation of the module has been initiated in FCE's Danbury, CT facility and testing up to 75% power level has been conducted successfully. As shown in Figure 6, the cell performance is fairly uniform when compared from stack to stack and within each individual stack. Greater than 65% stack module gross DC efficiency (based on lower heating value of NG) has been achieved.

Figure 5. 60 kW SOFC Stack Module: (left) During Assembly (before installation of Module Enclosure) and (right) After Installation in the Power Plant Test Facility

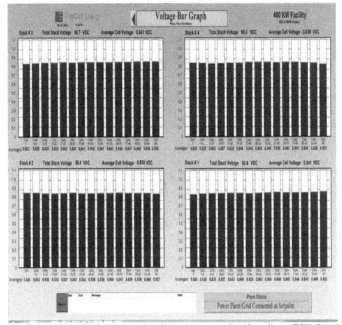

Figure 6. 60 kW SOFC Module Stack Performance Uniformity at 75% Load Level

SUMMARY

The IGFC Baseline Power Plant system, developed for central power generation, offers an electrical efficiency of 58.7% based on HHV of coal, while capturing > 99% of carbon (in the syngas) as CO_2. The NG-fueled 60 kW PCM system, suitable for dispersed power generation, has the potential to offer nearly 89% overall (thermal plus electrical) efficiency for CHP applications. Large-area (550 cm2 cell active area) SOFC stack building blocks with a cell count of 96, and 30 kW stack towers have been developed. A 60 kW SOFC module containing an assembly of identical stacks has been designed to lay the foundation for building blocks scalable for deployment in coal based power plants of the future. The hot tests have shown that the module and the non-repeat component designs (mechanical and electrical) meet the system requirements. Overall, the development of SOFC technology and systems will significantly advance the nation's energy security and independence interests (through effective utilization of the nation's vast coal reserves), address pollution and greenhouse gas concerns, and enhance the nation's economic growth.

ACKNOWLEDGEMENTS

This material is based upon work supported by the Department of Energy under Award Number DE-FC26-04NT41837. The guidance and helpful advice of DOE's Project Manager, Mr. Travis Shultz, is acknowledged.

REFERENCES

1.   H. Ghezel-Ayagh, S. Jolly, D. Patel, J. Nagar, K. Davis, C. Willman, C. Howard, M. Lukas, E. Tang, M. Pastula, and R. Petri, "SOFC Technology Development for Distributed and Centralized Power Generation", Abstract 291, *2012 Fuel Cell Seminar & Exposition, Uncasville, CT* (November 5-8, 2012)

2.   H. Ghezel-Ayagh, R. Way, P. Huang, J. Walzak, S. Jolly, D. Patel, C. Willman, K. Davis, M. Lukas, B. Borglum, E. Tang, M. Pastula, R. Petri, and M. Richards, "Solid Oxide Fuel Cell Stack Module Development for Fuel Flexible Power Generation", *ECS Transactions, 2011 Fuel Cell Seminar & Exposition, Orlando, FL*, Volume 42, Issue No. 1, pp. 233-239 (2012). Doe and R. Hill, *J. Electrochem. Soc.*, **152**, H1902 (2005).

3.   H. Ghezel-Ayagh," SOFC Development for Large-Scale Integrated Gasification Fuel Cell Power Plants", *International Colloquium on Environmentally Preferred Advanced Power Generation (ICEPAG), Costa Mesa, CA*, (February 7-9, 2012).

THE DEVELOPMENT OF PLASMA SPRAYED METAL-SUPPORTED SOLID OXIDE FUEL CELLS AT INSTITUTE OF NUCLEAR ENERGY RESEARCH

Chun-Liang Chang, Chang-Sing Hwang, Chun-Huang Tsai, Sheng-Hui Nien, Chih-Ming Chuang, Shih-Wei Cheng and Szu-Han Wu
Physics Division, Institute of Nuclear Energy Research
Taiwan, R.O.C.

ABSTRACT

Due to the advantages such as high thermal shock resistance and mechanical robustness in comparison with anode-supported cells (ASCs), the metal-supported cells (MSCs) attract more and more attention in the stationary and mobile applications. The planar MSCs composed of a well-prepared porous Ni-Fe plate as a supporting substrate, double layers of $La_{0.75}Sr_{0.25}Cr_{0.5}Mn_{0.5}O_{3-\delta}$ (LSCM) and nanostructured $La_{0.45}Ce_{0.55}O_{2-\delta}$/Ni (LDC-Ni) as an anode, a LDC layer as a diffusion barrier, a $La_{0.8}Sr_{0.2}Ga_{0.8}Mg_{0.2}O_{3-\delta}$ (LSGM) layer as an electrolyte, double layers of different contents of $Sm_{0.5}Sr_{0.5}CoO_{3-\delta}$-SDC (SSC-SDC) composite layers as a cathode were successfully fabricated by atmospheric plasma spraying (APS) technique at Institute of Nuclear Energy Research (INER). A post heat-treatment with a pressure of 0.8 kg/cm^2 at 850°C for 4 hrs was applied to the sprayed cells to improve its performances. The measured power densities of the MSC-MEA (Membrane Electrode Assembly) in cell test are 650, 568 and 443 mW/cm^2 at 750, 700 and 650°C, respectively. The durability tests of the MSC stack had performed for 1,100 hrs at the test condition of 400 mA/cm^2 and 700°C. The measured degradation rate is less than 1 %/kh, which indicates that the MSC-MEA made by INER reveals an inspired performance.

INTRODUCTION

A fuel cell is an electrochemical device which can directly convert the electrochemical energy of applied fuel to electricity. The solid oxide fuel cells (SOFCs) have some unique advantages over other types of fuel cell or traditional power generation technologies, including inherently high efficiency, low gas pollution and high fuel flexibility.[1,2] SOFCs with reduced operation temperatures (500-700°C) referred to intermediate temperature SOFCs (ITSOFCs) provide numerous advantages, such as the application of low-cost component materials, improvement of sealing capability, reduction of the interfacial reaction and chromium poisoning during cell operation. In order to reduce the operation temperatures of SOFCs, the technical development has focused on the development of both advanced materials and improved microstructures of IT-SOFCs components. Towards lowering operation temperatures, there is a tendency to shift ceramic-supported solid oxide fuel cells to metal-supported solid oxide fuel cells due to the potential benefits of low cost, high thermal shock resistance, mechanical robustness, better workability and quicker start-up.[3,4] In addition, the use of metallic substrates allows the use of convenient metal welding techniques for stack sealing and it can significantly reduce the manufacturing costs of SOFC stacks.

SOFCs are typically manufactured by using wet ceramic techniques such as tape casting combined with multi-step high temperature (up to 1400°C) sintering to obtain dense electrolytes. It is difficult to incorporate metallic substrates into the wet ceramic manufacturing processes without oxidizing the metallic substrates or significantly changing metallic substrate properties. Plasma spray processes as well as wet ceramic technique have the potential of low production cost due to their high deposition rates and ability to coat large surfaces.[5,6] In addition, plasma spray processes do not encounter substrate shrinkage problem and are favorable for the coating of substrates with reduced shrinkage as used in the metal-supported fuel cells. Besides the above advantages, plasma spray processes also show strong potential to enable the multi-layer fabrication of membrane electrode assemblies (MEA) on porous metallic substrates with sequential deposition steps.

Ishihara et al. and Goodenough et al.[7,8] had developed Sr- and Mg-doped $LaGaO_3$ (i.e. $La_{1-x}Sr_xGa_{1-y}Mg_yO_3$, LSGM) as an oxide ion conductor used as electrolyte for ITSOFCs. According to their researches, the ionic conductivity of $La_{0.8}Sr_{0.2}Ga_{0.8}Mg_{0.2}O_3$ is almost one order of magnitude larger than that of YSZ at the same temperature. LSGM also possesses a negligible electronic conduction below $800^{\circ}C$ and a stable performance over a long operating period of time.[9,10] Its superior ionic conductivity and chemical properties makes it a new generation electrolyte material for solid electrolyte of ITSOFC operated below 800°C.

In this report, the planar metal-supported solid oxide fuel cells (MSSOFCs) composed of a well-prepared porous Ni/Fe plate as a supporting substrate, double layers of $La_{0.75}Sr_{0.25}Cr_{0.5}Mn_{0.5}O_{3-\delta}$ (LSCM) and nanostructured $La_{0.45}Ce_{0.55}O_{2-\delta}/Ni$ (LDC/Ni) as an anode, a LDC layer as a diffusion barrier, a $La_{0.8}Sr_{0.2}Ga_{0.8}Mg_{0.2}O_{3-\delta}$ (LSGM) layer as an electrolyte, double layers of different ratios of $Sm_{0.5}Sr_{0.5}CoO_{3-\delta}$ (SSC) and $Sm_{0.15}Ce_{0.85}O_{2-\delta}$ (SDC) formed composite layers as a cathode were fabricated by atmospheric plasma spraying (APS) technique at Institute of Nuclear Energy Research (INER). These metal-supported cells (MSCs) are further assembled to become a short stack (single- or 5-cell stack) for testing their electrical performances.

EXPERIMENTAL

As shown in Figure 1, the APS system consisted of two plasma torches (a modified Praxair SG-100 dc torch with Mach I nozzle and a TriplexPro[TM]-200), two powder feeding systems, a cooling system, a furnace, an IR detector and a Fanuc Robot ARC Mate 120iB system to scan plasma torch. The dc plasma torch was operated at medium currents from 300 to 550 A and high voltages from 88 to 120 V. The mixed gas composed of argon, hydrogen and helium was used as plasma forming gas. A specially designed multi-gas mixer was applied to mix these gases uniformly. Argon, hydrogen and helium flow rates up to 60, 15 and 30 slpm, respectively, were controlled by using mass flow controllers. Details of experimental apparatus and typical plasma spraying parameters were given in another published paper.[11,12]

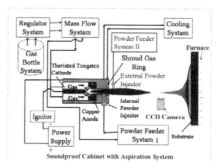

Figure 1. The scheme of APS set-up.

Commercial powders of LSCM, LDC/NiO/C, LDC, SSC/SDC and LSGM ($La_{0.8}Sr_{0.2}Ga_{0.8}Mg_{0.2}O_{3-\delta}$) were used in this study. These powders had average granule sizes from 20 to 50 μm. Carbon black of 15 wt% was applied as a pore former and it would be burnt out during the cell coating and annealing. The original particles of LDC and LDC/NiO agglomerated powders were less than 100 nm in size and the original particles of SSC agglomerated powders were between 200 nm and 400 nm in size. Layers of LSCM, LDC/NiO, LDC, LSGM, and SSC/SDC were plasma sprayed in sequence on the 10 x 10 cm^2 porous nickel-iron substrate to complete a multilayered MSC cell. The area of SSC/SDC layers was 72 cm^2 and the substrate with the permeability larger than 2 Darcy was 1.2 mm in thickness. A post heat-treatment with a pressure of 0.8 kg/cm^2 at 850°C for 4 hrs in air was applied to the sprayed cells to improve cell performance. The detailed information about the applied post heat-treatment can be found in our previous published paper.[13]

The electrical performances of MSC cells were performed in cell test and short-stack measurement configuration. The configurations of cell test and single-cell stack are showed in Figure 2 and 3, respectively. In Figure 2, the inert Al_2O_3 ceramic material is used for housing the tested cell. Platinum grids and leads at the anode and cathode side of the cell are used to measure cell current and voltage. Thermocouples close to the anode and cathode of cell are applied to measure temperatures on both anode and cathode sides. The MSC cell and the cell frame in Figure 3 are welded together by laser. The home-made glass-ceramic sealing materials that are not shown in Figure 2 are used to seal the tested single cell stack. Crofer 22 material is applied for making cell frames, anode plates and cathode plates. The $La_{0.67}Sr_{0.33}MnO_{3-\delta}$ (LSM) coating made by magnetron sputtering is applied for avoiding the chromium poison from contacts.

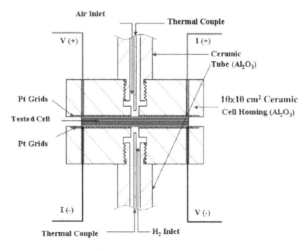

Figure 2. The scheme of MSC cell test housing for measuring MSC cell performance.

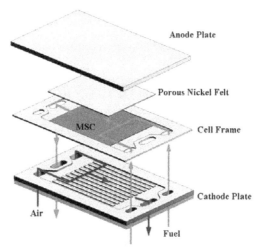

Figure 3. The 3D arrangement of components used in single-cell stack measurement configuration.

During the MSC cell test, the inert Al_2O_3 ceramic tubes are used to deliver hydrogen fuel with 800 ml/min and air oxidizer with 2000 ml/min to the tested cell. In the single-cell stack performance test, hydrogen and nitrogen gases with 800 and 200 ml/min flow rates were used as fuel gas, and the air oxidizer was fed with 2,000 ml/min flow rate. During long term durability test of the single-cell stack, the applied current density and operating temperature are kept at 400 mA/cm^2 and 700°C, respectively.

During 5-cell stack performance test, the reformed gas with 6,820 sccm flow rate and air with 13,000 sccm rate were respectively used as fuel and oxidizer gas to complete I-V-P and durability tests. The reformed gas generated from methane and water vapor through the home-made reformer installed in the SOFC measurement system. According to our experiences about measuring the degradation rate of SOFC MSC cell stack, the most suitable applied current density is located around 400 mA/cm^2 at 700℃. So, the constant current density is kept at 400 mA/cm^2 in long term durability test.

Solartron 1255 and Solartron 1287 are applied for the AC impedance measurement, and Prodigit 3356 DC electronic load is applied for the power measurement. Platinum grids and leads at the anode and cathode side of the cell are used to measure cell current and voltage.

RESULTS AND DISSCUSION

A typical microstructure of an APS prepared cell after test is shown in Figure 4. Six-functional layers of this cell were deposited on the porous nickel-iron substrate by atmospheric plasma spraying processes. The LSGM layer is found to be quite dense with closed pores and without cracking through. Good interfacial adhesion is found between different functional layers of a cell. Figure 5 gives the SEM and TEM image of nanostructured LDC/Ni anode in a high magnification. This anode contains nano pores, pores larger than 100nm, nano Ni particles, nano LDC particles and clusters formed by nano Ni particles and nano LDC particles.

Figure 4. Cross sectional SEM micrograph of APS cell after hydrogen reduction

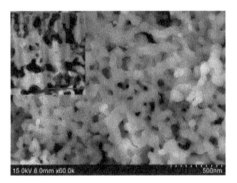

Figure 5. SEM and TEM image of nano-structured LDC/Ni anode.

The current-voltage (I-V) and current-power (I-P) curve of the single MSC cell at 750, 700 and 650°C are shown in Figure 6. The high open-circuit voltages (above 1.0 V) indicate that the LSGM electrolyte is dense enough so that gases permeated through it are negligible. The maximum power densities were 650, 568 and 443 mW/cm² at 750, 700 and 650°C, respectively. Figure 7 shows the dependences of the cell voltage and power density of the MSC cell tested in single-cell stack configuration as a function of the current density at temperatures from 650 to 750°C. The maximum power densities are 391, 306 and 194 mW/cm² at 750, 700 and 650°C, respectively. Figure 8 gives the long term durability test curves of single-cell MSC stack testing at 400 mA/cm² constant current density and 700°C operating temperature for the time period of 1,100 hours. Figure 8 shows that single-cell MSC stack operated at 700°C could deliver about 301 mW/cm² power density, and after applying 400 mA/cm² constant current density, the cell voltage starts to increase slightly and reach a maximum value of 0.752 V, then decreases to 0.750 V at the end of this test. According to Eq. (1), the degradation rate of single-cell MSC stack is slightly lower than 0.3%/kh, which is estimated to ~0.3 %/kh.

$$degradation\ rate\ (\%/kh) = \frac{V_i - V_f}{V_i t} \times 100000 \qquad ...Eq.\ (1)$$

In Eq. (1), the V_i, V_f and t refer to initial cell voltage, final cell voltage and operating time (in hours) in long term durability test.

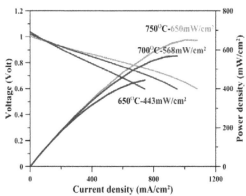

Figure 6. The I-V-P curves of single MSC cell with hydrogen at 650, 700 and 750°C, respectively.

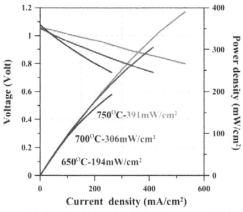

Figure 7. The I-V-P curves of single-cell MSC stack with hydrogen at 650, 700 and 750°C, respectively.

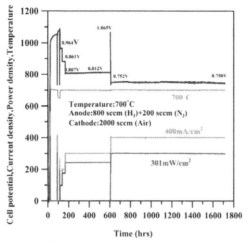

Figure 8. Long-term durability test results of single-cell MSC stack with hydrogen.

Figure 9 shows the I-V-P results of 5-cell MSC stack testing at 700°C and using reformed gas as fuel in home-made SOFC power generation system constructed by Institute of Nuclear Energy Research (INER). The average OCV of MSC cell is about 0.98 V, which is slightly lower than that (about 1.05 V) of single-cell MSC stack when 80% hydrogen was used as fuel. This difference is due to the composition of reformed gas, which has only 62.2% hydrogen. Moreover, the other compositions of the reformed gas are carbon monoxide and auxiliary gas (including nitrogen and residual methane) with about 18.7 and 19.1%, respectively. The total output power of 5-cell MSC stack could reach 115.4 Watts at 700°C and 0.77 V of average cell voltage. The durability results of 5-cell MSC stack under the testing condition of 700°C and 29 A (equate to 400 mA/cm^2) are shown in Figure 10. The stack voltages oscillated in the range of 3,650~3,950 mV which might be due to the complex gas interactions between residual methane, water vapor, CO and CO$_2$ in anode side of MSC cell. From Figure 10, it can be found that the initial output power of the 5-cell MSC stack at 29 A was about 113 Watts. After 110 hrs of durability operation, the stack power still keeps at 113 Watts, which results in no degradation of the MSC stack. These results reveal that the MSC cells made by APS technique with a SSC/SDC cathode layer have impressive power density and durability performances at 700°C, whether the fuel gas is mainly composed of hydrogen or reformed gas.

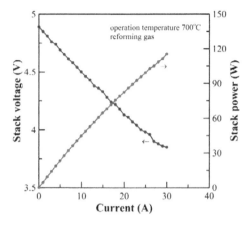

Figure 9. I-V-P curves of 5-cell MSC stack with reformed gas at 700°C.

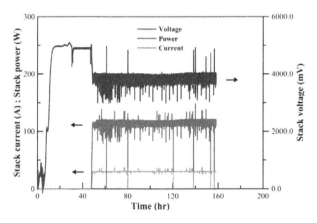

Figure 10. Durability test results of 5-cell MSC stack with reformed gas at 700°C.

CONCLUSION
The planar MSCs composed of a well-prepared porous Ni/Fe plate as a supporting substrate, double layers of LSCM and nanostructured LDC/Ni as an anode, a LDC layer as a diffusion barrier, a LSGM layer as an electrolyte, double layers of different ratios of SSC and SDC formed layers as a cathode were successfully fabricated by APS technique at INER. The measured power densities of the

MSC-MEA (Membrane Electrode Assembly) in cell test are 650, 568 and 443 mW/cm^2 at 750, 700 and 650°C, respectively. For single cell MSC stack with 80% hydrogen, the maximum power densities are 391, 306 and 194 mW/cm^2 at 750, 700 and 650°C, respectively, and the estimated degradation rate of ~0.3%/kh is obtained after 1,100 hrs operation. For 5-cell MSC stack with reformed gas, the stack delivers 115.4 Watts at 700°C and 0.77 V of average cell voltage and shows no degradation after 110 hrs operation.

REFERENCES
[1]T. H. Etsell and S. N. Flengas, Electrical Properties of Solid Oxide Electrolytes, *Chem. Rev.*, **70**(3), 339-376 (1970).
[2]H. Inaba and H. Tagawa, Ceria-based Solid Electrolytes, *Solid State Ion.*, **83**, 1-16 (1996).
[3]Z. Wang, J. O. Berghaus, S. Yick, C. Decès-Petit, W. Qu, R. Hui, R. Maric and D. Ghosh, Dynamic Evaluation of Low-temperature Metal-supported Solid Oxide Fuel Cell Oriented to Auxiliary Power Units, *J. Power Sources*, **176**, 90-95 (2008).
[4]M. C. Tucker, G. Y. Lau, C. P. Jacobson, L. C. DeJonghe and S. J. Visco, Stability and Robustness of Metal-supported SOFCs, *J. Power Sources*, **175**, 447-451 (2008).
[5]R. Hui, Z. Wang, O. Kesler, L. Rose, J. Jankovic, S. Yick, R. Maric and D. Ghosh, Thermal Plasma Spraying for SOFCs: Applications, Potential Advantages, and Challenges, *J. Power Sources*, **170**, 308-323 (2007).
[6]R. Henne, Solid Oxide Fuel Cells: A Challenge for Plasma Deposition Processes, *J. Therm. Spray Technol.*, **16**(3), 381-403 (2007).
[7]T. Ishihara, H. Matsuda and Y. Takita, Doped LaGaO3 Perovskite Type Oxide as a New Oxide Ionic Conductor, *J. Am. Chem. Soc.*, **116**, 3801-3804 (1994).
[8]K. Huang, R. S. Tichy and J.B. Goodenough, Superior Perovskite Oxide-Ion Conductor; Strontium- and Magnesium-Doped LaGaO$_3$: I, Phase Relationships and Electrical Properties, *J. Am. Ceram. Soc.*, **81**, 2565-2575 (1998).
[9]T. Inagaki, F. Nishiwaki, S. Yamasaki, T. Akbay and K. Hosoi, Intermediate Temperature Solid Oxide Fuel Cell based on Lanthanum Gallate Electrolyte, *J. Power Sources*, **181**, 274-280 (2008).
[10]C.-s. Hwang, C.-H. Tsai, J.-F. Yu, C.-L. Chang, C.-M. Lin, Y.-H. Shiu and S.-W. Cheng, High Performance Metal-supported Intermediate Temperature Solid Oxide Fuel Cells Fabricated by Atmospheric Plasma Spraying, *J. Power Sources*, **196**, 1932-1939 (2011).
[11]C. S. Hwang, C. H. Tsai, C. H. Lo and C. H. Sun, Plasma Sprayed Metal Supported YSZ/Ni–LSGM–LSCF ITSOFC with Nanostructured Anode, *J. Power Sources*, **180**, 132–142 (2008).
[12]C.H. Lo, C.H. Tsai and C.S. Hwang, Plasma-Sprayed YSZ/Ni-LSGM-LSCo Intermediate-Temperature Solid Oxide Fuel Cells, *Int. J. Appl. Ceram. Technol.*, **6**(4), 513-524 (2009).
[13]C.H. Tsai, C. S. Hwang, C. L. Chang, J. F. Yu and S. H. Nien, Post-heat treatment pressure effect on performances of metal-supported solid oxide fuel cells fabricated by atmospheric plasma spraying, **197**, 145-153 (2012).

DEVELOPMENT AND APPLICATION OF SOFC-MEA TECHNOLOGY AT INER

Maw-Chwain Lee*, Tai-Nan Lin, Ruey-yi Lee
Institute of Nuclear Energy Research
Longtan Township/Taiwan (R.O.C.)
Tel.: +886-3-471-1400 Ext. 5930
Fax: +886-3-471-1411
mclee@iner.gov.tw

ABSTRACT

The Institute of Nuclear Energy Research (INER) initiated the development of the solid oxide fuel cell membrane-electrode-assembly (SOFC-MEA) technology in 2003. Now, substantial progresses have been achieved on the related techniques. Fabrication processes for planar anode/electrolyte-supported-cell (ASC/ESC) by conventional methods and metallic-supported-cell (MSC) by atmospheric plasma spraying are well established. At this stage, the maximum power densities of INER's ASCs are 652 mW/cm^2 at 800 °C for intermediate-temperature SOFC (IT-SOFC) and 608 mW/cm^2 at 650 °C for low-temperature SOFC (LT-SOFC). The power densities of INER's MSCs are 540 and 473 mW/cm^2 at 0.7 V and 700 °C for a cell and a stack tests, respectively. Durability tests for ASC/MSC at constant current densities of 300/400 mA/cm^2 indicate that the degradation rates are less than 1%/khr. Comparable or higher performance is now achieved with respect to the commercial cells. Innovative materials, structures, and the fabrication process for SOFC-MEA are improved to upgrade the MEA qualities and enhance the cost reduction. The diverse SOFC-MEAs are developed for specific issues to simultaneously solve the problems of energy and CO_2 greenhouse effect. A 1-kW SOFC power system is designated and a pilot-scale production line of SOFC-MEA is constructed and operated at INER. Efforts are continuously input to solve the fatal problems, select the best type of SOFC-MEA, and assure the right choice for commercialization of the SOFC industry.

INTRODUCTION

The lack of indigenous energy resources and high dependence on energy imports (~99%) indicate that the energy security is crucial in Taiwan[1]. Mikkelsen *et al.* created the term of "The teraton challenge "and described that the increase in atmospheric carbon dioxide is linked to climate changes[2]. Fig. 1 shows the fatal problem of carbon cycle essentially triggered by the excess emission of CO_2 from the power generation. It is an urgent need to reduce the accumulation of CO_2 in the atmosphere. As a member of this global village, Taiwan has committed itself to fulfil the obligation of common, share the related responsibility with regard to the reduction of greenhouse gas (GHG) emissions, and to sustain the progress of national

41

economic growth. As a result of the nuclear accidents at Fukushima in Japan, a newly announced nuclear energy policy promises eventually to make the island 'nuclear free' and no life extensions will be granted to the existing nuclear power plants in Taiwan. The solutions of impacts between the national electricity shortages and revised energy policy, while keeping benign or moderate growth of economy and continuing reduction of carbon dioxide emissions to meet the international goals, are being comprehensively pursued. Nowadays, the SOFC technology, with the main features of high fuel flexibility, modular design, less pollutant emissions and high energy conversion efficiency, is considered as a promising clean technology beneficial to resolve the dilemma of economic growth, energy security and environmental protection[1]. In compliance with the national energy policy, the SOFC project at INER focuses on establishing manufacture capability of SOFC-MEA components and integration technology of SOFC power system. Since the commitment to developing the SOFC technology in 2003, this institute has set its short-term target of 1~5 kW SOFC distributed power generation systems and will then extend its long-term prospect to integration with the integrated gasification combined cycle (IGCC) technology for biomass and coal based central power generation and large demonstration systems[1]. To accelerate the progress on SOFC development, three main sub-projects are formed, i.e., MEA development, stack development, and power system development. SOFC-MEA sub-project is the most critical issue and has its main tasks to identify the key technical scopes and then to find out proper resolutions. Meanwhile, it should intensively incorporate with other sub-projects to clarify interfaces and avoid erroneous discrepancy among different groups.

After years of elaborate efforts, INER has now possessed the critical core technologies from powder to power for a distributed kilowatt SOFC power system. On the basis of current progress, a national SOFC technology roadmap from now to 2030 is proposed, where the operating temperatures, power densities, energy conversion and system efficiencies, degradation rates or durability, cost and power levels are discreetly projected. The progressive achievements in the past few years on the development and application of INER-SOFC-MEA technology are briefly described.

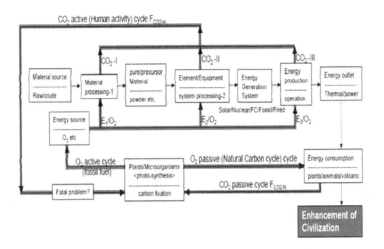

1. $|F_{CO2-H}-F_{CO2-N}|=3.9\% \ F_{CO2-N}$ (F=annual carbon-flow) : E_i=energy input at i-th step
2. $[CO_2(270ppm/1810) \rightarrow 200yrs \rightarrow CO_2(385ppm/2010) \rightarrow CO_2(570ppm/2100)$ (conc. of atmospheric CO_2) $\rightarrow$ climate change $\rightarrow$ Global Temperature, as $\Delta T=1.9°C$, Δh=mean sea level increase=1m [2]
3. Energy Generation System $\rightarrow$ Solar/ Nuclear/ wind/ Fossil fuel/ FC etc.

Fig. 1 The fatal problem of Carbon Cycle

EXPERIMENTAL

1. Development of SOFC-MEA technology

SOFCs, which are well recognized as environmentally friendly, highly efficient devices, and alternatives to conventional energy conversion systems, are considered to be one of the most important power generation technologies of the future[3]. A typical MEA for SOFCs consists of two electrodes (anode/cathode) separated by an electrolyte. Hence, key issues for development of SOFC-MEA technology include: 1) development of SOFC-MEA materials used for anode, cathode, and electrolyte layers, 2) development of SOFC-MEA structure for selection of product orientation, 3) development of fabrication process of SOFC-MEA to manufacturing the product, and 4) development of cell test station of SOFC-MEA for MEA performance test and identification. These issues are described as follows.

1.1 Development of SOFC-MEA materials: The planer type SOFCs with ASC and MSC are essentially selected for operation at temperature range of 400 to 850 °C at INER. Key materials for anode, electrolyte, and cathode are $NiO + Y_{0.08}Zr_{0.92}O_{2-\delta}$ (8YSZ), $NiO + Sm_xCe_{1-x}O_{2-\delta}$ (SDC), $NiO + Gd_xCe_{1-x}O_{2-\delta}$ (GDC), $NiO + La_{1-x}Sr_xGa_{1-y}Mg_yO_{3-\delta}$ (LSGM); 8YSZ, GDC, SDC, LSGM,

$BaY_xCe_yZr_{1-x-y}O_{3-\delta}$ (BYCZ); and $La_{1-x}Sr_xMnO_{3-\delta}$ (LSM), $La_{1-x}Sr_xCo_yFe_{1-y}O_{3-\delta}$ (LSCF), $Ba_{1-x}Sr_xCo_yFe_{1-y}O_{3-\delta}$ (BSCF), $Sm_xSr_{1-x}CoO_{3-\delta}$ (SSC), $Sm_1Ba_xSr_{1-x}Co2O_{5+\delta}$ (SBSC); respectively[1]. The nano-scale powders of YSZ / LSGM / SDC / GDC / pyrochlore are prepared and processed as well as used for fabrication of SOFC-MEA with thin (< 10 μm) and fully dense film of electrolyte to achieve high power density. Table 1 shows the general / effective materials used in SOFC-MEA. The characteristics of SOFC-MEA related ceramic materials / powders are different with respect to the usage of anode, cathode, and electrolyte. The process for powder preparation decides the characteristics of its product. Hence, the selection of powder preparation process is critical for the development of SOFC-MEA materials[3]. Sunarsoa et al.[4] and Kharton et al.[5] reported that at similar operating conditions, different preparation methods for a particular ceramic membrane with certain material composition resulted in a different value of the oxygen flux/electrical transport properties and oxygen permeation. This is mainly caused by the effect of the unique microstructure achieved from each method, which impacts upon its physical and transport properties. In general, about five synthesis methods have been employed quite extensively to prepare mixed oxide compounds at INER. Table 2 shows the main processes employed for preparation of fine ceramic powders.

Table-1 Typical materials used in various types of SOFC-MEA

Types MEA	ESC	ASC-I	ASC-II
Anode	NiO (30~50wt%) +8YSZ th = 50 μm	NiO +8YSZ th = 600 μm (Anode substrate)	NiO +GDC/SDC th => 600 μm
Electrolyte	8YSZ th = 140 μm (Electrolyte substrate)	8YSZ th = 3~10 μm	LSGM/ LSGMC /GDC/ SDC etc. th =7~15 μm
Cathode	LSM th = 50 μm	LSM th = 30~60 μm	LSCF-GDC/ LSM /SSC / LNF etc. th = 30~60 μm
Operating Temp. (ºC)	1000~800	800~600	650~450

Table-2 Main processes for preparation of fine ceramic powders used in SOFC-MEA

Item	Contents	Process
1.	Electrolyte: 8YSZ/GDC/YDC/LSGM/SDC/ LSGMC	a. Hydrothermal
		b. Co-precipitation
2.	Anode: NiO+8YSZ/GDC/SDC/LSGM	c. Sol-gel
		d. GNC Process
3.	Cathode: LSM/LSCF/BSCF/BSSC/BSAF/SSC	e. Physical (Solid-state reaction)
4.	Nano- / Submicron-Scale Ceramic Powders	f. Innovative Processes (Spray Pyrolysis etc.)

Advantages and drawbacks are inherent to the process itself. The requested characteristics of fine ceramic powders used for electrode fabrication of anode and cathode are less strict than those of electrolyte. The hydrothermal method and glycine-nitrate combustion process (GNP) are essentially employed to prepare the nano-scale ceramic powders at INER[6-9]. The solid state reaction method may be used for preparation of the mixed ceramic powders without the restrict limitations of the characteristics of particle size and surface area.

A. Anode Materials:

NiO+8YSZ / NiO+SDC / NiO+GDC / NiO+LSGM / NiO+Pyrochloes are key materials used for preparation of anode electrodes. The weight ratio of NiO / electrolyte material is near 50 % (or in the range of 40~70 %) to obtain a compromise between electron / oxygen ion conductivity and mechanical strength for an ASC/MSC/ESC[6-9]. The functional materials, such as Al_2O_3, are additives of anode materials as the reagent for sintering enhancement to obtain an anode substrate with high mechanical strength[10]. A homogeneous anode powder with micro / submicron-scale particle size is generally accepted for fabrication of an anode substrate. However, the characterization task must be well done and the characteristics of anode powder must be strictly controlled to assure the powder quality to be committed with the requirements of the fabrication process for anode layer/substrate. Key characteristics of the anode powder are described in Table 3.

B. Electrolyte Materials

8YSZ / 4YSZ / $Sc_xZr_{1-x}O_{2-\delta}$ (ScSZ) / $[(ZrO_2)_{1-y}(CeO_2)_y]_{1-x}(Sc_2O_3)_x$ (SCSZ), and GDC / SDC / LSGM / pyrochlores are key electrolyte materials employed for fabrication of HT-SOFC,

IT-SOFC, and LT-SOFC, respectively. BYCZ is employed as the electrolyte for proton-conducting SOFC (p-SOFC) and operated at the temperature range of 350 ~ 500 °C[11-13]. The characteristics of electrolyte powders are more strictly required and controlled than those of electrode materials. Essentially, the electrolyte layer must be satisfied the requirements of 1) thin film(thickness less than 10 μm for ASC and MSC) to increase the transport rate and reduce the resistance of the oxygen ion conductivity across the electrolyte layer, 2) extremely high sintering density (near 100 % of theoretically density) to assure the gas tight and achieve the theoretical value of open circuit voltage (OCV) of a cell, 3) thin film(thickness near 150 ~ 200 μm for ESC to increase the transport rate and reduce the resistance of the oxygen ion conduction across the electrolyte layer as well as to equip with the sufficiently mechanical strength for MEA support. A high densification of the sintering elements is required for a material to act as an electrolyte. The powder precursors are believed to have a great effect on the densification process, including the sintering temperature. Nano-scale particles tend to show better sintering ability than larger particles. Low temperature processing is considered critical for energy-saving and cost reduction. YSZ is considered to be one of the most reliable candidates for the electrolyte and has been widely used. However, the high operating temperature of traditional SOFCs with YSZ electrolytes results in high costs as well as problems with materials selection and degradation. Therefore, the development of LT-SOFCs (400 - 600 °C) is critical for cost reduction, long-term performance stability, and materials selection. To reduce the operating temperature, two approaches are widely applied to lower the resistance of the dense electrolyte membranes, including decreasing the thickness of the traditional YSZ electrolyte or using alternative materials with higher ionic conductivities at low temperatures[3]. SDC / GDC have attracted much attention in recent years, especially as a possible solid electrolyte substitute for YSZ. Their superior oxygen ionic conductivity[3] and good compatibility with electrodes allow the related SOFCs to have lower operating temperatures (from 1000 °C down to 600 ~ 800 °C). LSGM has advantage of high oxygen ionic conductivity at the temperature range of 550 ~ 800 °C and is used as one of electrolytes of IT-SOFC and LT-SOFC[14]. However, it owns the drawbacks of chemical instability due to the gradient/element interactions between electrolyte and electrode. Composite electrolyte based on the materials of ScSZ / SCSZ / YSZ has been proposed and patented for HT-SOFC and IT-SOFC to improve the mechanical strength and durability as well as enhance the oxygen ion conductivity at low temperature (700~850°C)[15-16]. The composite electrolyte may comprise sintered mixture of ScSZ/SCSZ and YSZ. Pyrochlore materials are the other candidates to be an electrolyte of LT-SOFC[17]. However, more efforts must be input to develop the related techniques. Table 4 describes the summarized contents of this issue. A high densification of the sintering elements is required for a material to act as an electrolyte. The powder precursors are believed to have a great effect on the densification process, including the sintering temperature. Hence, the interaction between processes of the material preparation and element fabrication of an excellent electrolyte is critical. Preparation of dense electrolyte films on porous electrodes or substrates is the critical step in the fabrication of high-performance solid-state fuel cells. Key processes employed to prepare these thin-film electrolytes on porous substrates include three categories: 1) vapor-phase deposition methods, such as spray pyrolysis, plasma coating, and pulsed laser deposition[3, 14], 2) liquid-phase deposition methods, such as spin coating, sol-gel, and the liquid-state deposition method, and 3) thin film techniques for the chemical engineering process with particle deposition / consolidation methods, such as dry pressing, tape casting, screen printing, and slurry spin coating .The process selection is quite important in this regard. With liquid-phase deposition techniques, especially with the spin coating of a precursor, compositional homogeneity at the molecular level can be achieved with a precursor solution rather than a prefabricated powder. This process favours the fabrication of an extremely thin ceramic film and is also relatively inexpensive in terms of equipment and processing costs[3].

Table.3 Key characteristics of materials used for SOFC-MEA

Particle Size	DLS
Surface Area and Pore size	Surface area & Pore size analyzer
Element	EDS, XPS
Crystal phase	XRD
Thermo-Behavior analysis	TGA-DTA
Density	Ultrapycnometer
Microstructure	SEM, TEM
Formulation stability	Zata Potential

Table.4 Key electrolytes materials of SOFC-MEA

Types / MEA	HT-SOFC	IT-SOFC	LT-SOFC
Electrolyte	YSZ/SCSZ/ ScSZ	8YSZ	LSGM/ LSGMC /SDC/GDC/SSZ/ Pyrochlore etc.
Operating Temp. (°C)	1000~800	750~600	650~400

Note: Key issues 1.Oxygen ion conductivity
                2.Thermal expansion coefficient
                3.Crystal phase

C. Cathode Materials

LSM/LSCF/BSCF/SSC/SBSC are key cathode materials developed and used for SOFC-MEA at this stage. Many new materials in this regard are still endeavoured for improvement of the quality of SOFC[3, 18-23]. The characteristics of cathode ceramic powders are not strictly requested because the cathode layer must be porous and the fabrication process is not so complicated. Hence, the ceramic powders of micro-/sub-micro-scale particle size are good for cathode materials. Key characteristics of a good cathode material with enhanced electrochemical performance include the following: (1) a high rate of oxygen diffusion through the material to assure the rapid diffusion of oxygen; (2) a high electronic conductivity; (3) a high oxygen ion conductivity; (4) a high catalytic activity for the reduction of oxygen; and (5) a suitable thermal-expansion coefficient (TEC) to assure chemical and mechanical compatibility between electrode[8, 19]. LSM/LSCF are proven and commercially employed as cathode materials. The cobalt- or iron–cobalt-based perovskites are known to have better performance than that of the typical LSM-YSZ composite as cathode materials. LSMs have been used extensively as cathode materials in YSZ-based SOFCs due to their matching TEC, assuring mechanical compatibility and high electronic conductivity. LSCs and their ion-substituted derivatives LSCFs are known to have high catalytic activity for the reduction of oxygen and high electronic conductivity. However, the large TEC of LSC results in the problem of mechanical compatibility. The Co-based perovskite material SSC also has high electronic conductivity[8, 19]. BSCF gives excellent performance as the cathode for a reduced-temperature (600 °C) SOFC due to its high rate of oxygen diffusion through the material. The oxygen-reduction in BSCF occurs not only on the triple-phase boundary (TPB), but also proceeds on the surface of the electrode. However, the BSCF cathode is not compatible with zirconia-based electrolyte because of their chemical interaction to produce an interfacial-insulating layer and reduce the catalytic activity of the electrode and the issue of mismatch of TEC[8]. This disadvantage can be overcome by the addition of GDC to BSCF to adjust the TEC of the BSCF cathode. Chang et al.[19] investigated the characterization of anode-supported solid oxide fuel cells with composite LSM-YSZ and LSM-GDC cathodes. It shows that the composite cathode layer of GDC-LSM/LSM is a better choice as the advanced cathode for the SOFC operated at the lower temperature (near 700 °C). To improve the rate of the oxygen reduction in the cathode and enhance the performance of the cell, the fabrication work of the cathode electrode must be concentrated on the targets of the increase of the TPB length and the selection of the suitable cathode materials to satisfy the requirements of the chemical compatibility and the match of the TEC. GDC is known to be chemically compatible with the BSCF and other cathode/electrolyte materials, and possesses a higher ionic conductivity than that of the YSZ electrolyte[20]. Therefore, the GDC can be coated onto an YSZ/SDC/LSGM electrolytes as a barrier layer to prevent the direct interaction between the cathode and electrolytes. In addition, GDC is also a modulator for adjustment of TEC of cathode materials and solve the problem of the mismatch of the TEC. Composite cathode layers of electrolyte-cathode/cathode are strongly recommended to simultaneously solve the problems of chemical and mechanical compatibilities between cathode and electrolyte.

1-2. Development of SOFC-MEA Structure

The planer type SOFCs with ASC and MSC are selected for operation at temperature range of 400 to 850 °C. ESC and Cathode Supported Cell (CSC) are also concerned in our scope of research for consideration of their special characteristics and functions as well as niche benefits, such as the anti-redox capability of ECS and cost reduction of CSC.

1-3 Development of fabrication process of SOFC-MEA

The Fabrication process of SOFC-MEA developed at INER is shown in Fig.2 and Tables 5A/5B as well as intensively described in the previous papers[3, 6-10, 19-20] and patents[9, 18, 21, 24-26]. This process is developed to cover the related technologies of fabrication of SOFC-MEA from powder to power. The matches between materials and fabrication process are critical to assure the great achievements of the cell quality and cell performance. The yield of this process is approximately 100% so that the cost reduction is possible. Further improvement of this process is to reduce the energy consumption and simplify the number of sintering steps. Key achievements in the development of the entire content of SOFC-MEA (ASC) at INER are described as follows:

A. HT-SOFC-MEA (700-1000 °C)

The typical SOFC-MEA of HT-SOFC is with the structure of NiO+YSZ ‖ YSZ ‖ composite cathode (e g. YSZ-LSM/LSM, GDC-LSM/LSM, SDC/SSC-SDC, etc.). The microstructure and the results of cell performance are shown in Figs. 3 and 4. The MPD is near 400 W/cm^2 (at 800 °C) and the degradation rate is near 2.3 % / Khr with two thermal cyclings and 7,000 hours operation[3, 26].The updated degradation rate is 0.38 % / Khr with five thermal cyclings and 7,000 hours operation, see Fig. 5A. The performance result of double cell stack is shown in Fig. 5B. It indicates that the power rate is over 25 W/ cell (at 800 °C).

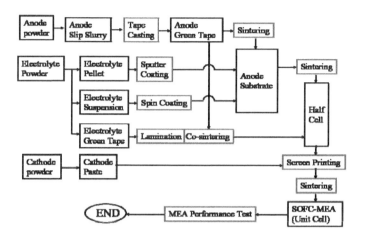

Fig-2 Fabrication process of SOFC-MEA at INER

Table 5A. Main compositions and thicknesses for three types of INER-SOFC-MEAS

Temperature(°C)	Composition			thickness (um)		
	cathode	electrode	anode	cathode	electrode	anode
700-1000 (HT)	YSZ+LSM/LSM	YSZ	NiO+YSZ	25	7	800
600-800 (IT)	GDC-LSM/LSM, SDC/SSC-SDC SDC/SDC-SBSC	YSZ	NiO+YSZ	39	7.5	650
400-650 (LT)	SSC-SDC/SSC	SDC	NiO+SDC	45	15	900

Mechanical strength~143 MPa
Anode permeability ~ 0.016 Darcy

Table 5B. Power performance for three types of INER-SOFC-MEAS

Temperature(°C)	OCV (V)	MPD (mW/cm²)	Degradation(%/khr)
700-1000 (HT)	1.074	490 (900°C)	2.3→3.4 (7800hr)
600-800 (IT)	1.009	559 (800°C)	...
400-650 (LT)	0.882	607.68 (650°C)	3.0 (950hr)

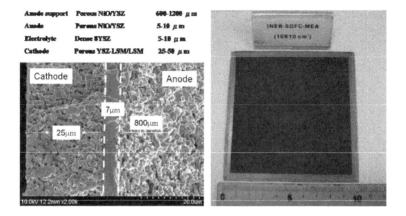

Anode support	Porous NiO/YSZ	600-1200 μ m
Anode	Porous NiO/YSZ	5-10 μ m
Electrolyte	Dense 8YSZ	5-10 μ m
Cathode	Porous YSZ-LSM/LSM	25-50 μ m

Fig-3 The typical microstructure / specification of INER-HT-SOFC-MEA

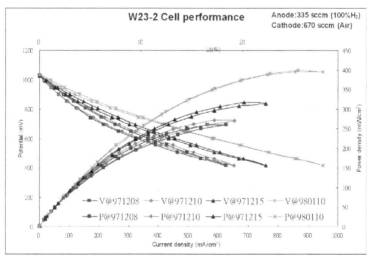

Fig-4 V-I-P performance test of INER-HT-SOFC-MEA-W23-2

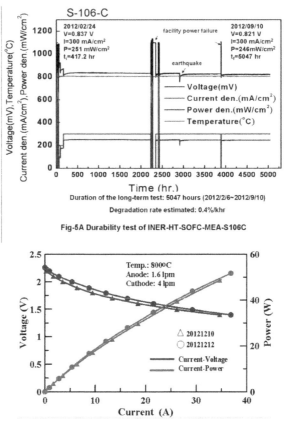

Fig-5A Durability test of INER-HT-SOFC-MEA-S106C

Fig.5B V-I-P of performance test for double-cell stack

B. IT-SOFC-MEA (600 ~ 800 °C)

The typical SOFC-MEA of IT-SOFC is with the structure of NiO+YSZ ∥ YSZ ∥ composite cathode (e g. SDC/SDC-SBSC, etc.). The microstructure and the results of cell performance are shown in Figs. 6 & 7. The maximum power density (MPD) is 652 mW/cm^2 (at 800 °C, OCV = 1.099 V) and the degradation rate is near 1.0 % / Khr with 1,400 hour operation[3, 23].

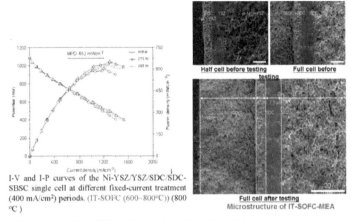

I-V and I-P curves of the Ni-YSZ/YSZ/SDC/SDC-SBSC single cell at different fixed-current treatment (400 mA/cm²) periods. (IT-SOFC (600~800°C)) (800 °C)

Half cell before testing   Full cell before testing

Full cell after testing
Microstructure of IT-SOFC-MEA

IT(600~800°C) SOFC-MEA with P_max =652 mW/cm² (800°C)

Fig .6 INER- IT-SOFC (600~800°C)

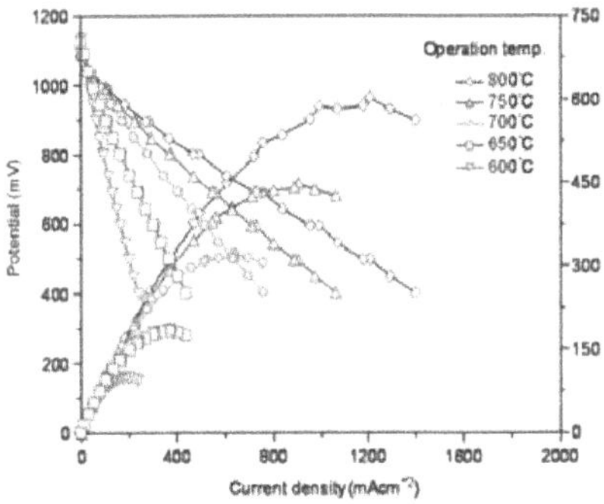

I-V and I-P curves of the Ni-YSZ/YSZ/SDC/SDC-SBSC single cell at different operation temperatures (IT-SOFC(600-800 C)).

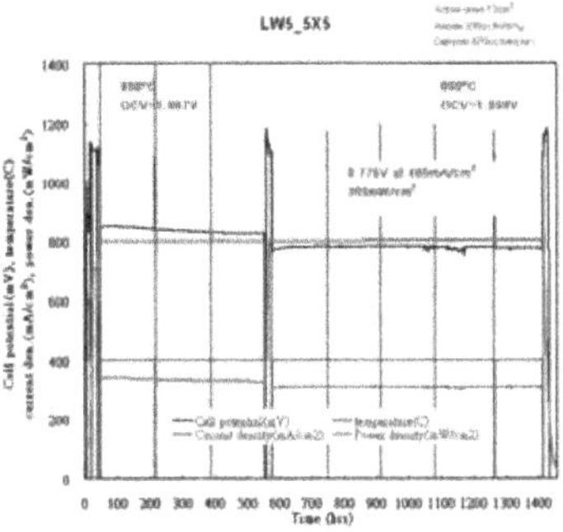

Durability test ~1400hrs

Fig .7 Performance test of INER- IT-SOFC (600~800°C)

C. LT-SOFC-MEA (400 ~ 650 °C)

The typical SOFC-MEA of IT-SOFC is with the structure of NiO+SDC ‖ SDC ‖ composite cathode (e g. SSC-SDC/SSC, etc.). The microstructure and the results of cell performance are shown in Figs. 8 & 9. The MPD is 607.68 mW/cm² (at 650 °C) and the degradation rate is near 3.0%/Khr with 950 hours operation[1, 3]. The redox problem of SDC electrolyte has been identified[3, 27] and the scheme to solve this issue will be proposed in the near future.   LT-SOFC should be a technology of choice for these applications as long as we are in a hydrocarbon-based energy infrastructure[28]. SOFCs based on the proton conducting BZCY electrolyte were also studied and prepared for the development of LT-SOFC[11-13].

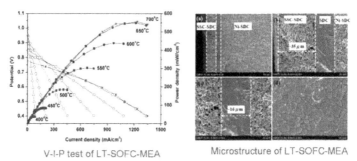

V-I-P test of LT-SOFC-MEA   Microstructure of LT-SOFC-MEA

Fig. 8 INER- LT-SOFC (400~650°C) (Ni-SDC/SDC/SDC-SSC/SSC)

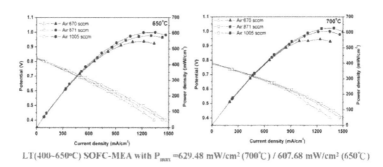

LT(400~650°C) SOFC-MEA with P_{max} =629.48 mW/cm² (700°C) / 607.68 mW/cm² (650°C)

Fig.9 INER- LT-SOFC (400~650oC) (Ni-SDC/SDC/SDC-SSC/SSC)

1-4. Development of Cell Test Station of SOFC-MEA

The cell performance of SOFC-MEA must be done in the cell test station. Both cell test stations with open or close systems are employed at INER. Cell test station with close system includes the types of ProboStat Unit and single cell stack equipped with on-line Micro-GC for gradient/composition analysis of the outlet gas streams. Either CH_4 or H_2 are employed as fuels and air is used as oxidant. Both Characteristics of cell performance and the related techniques for enhancement of SOFC-stack fabrication are investigated. Key issues of our achievement in this regard are :A) finding of the breeding phenomenon of nickel in anode of solid oxide fuel cell via electrochemical reaction to identify the innovative mechanism and behaviour of electro-chemical reaction[29]: The breeding of nano-scale catalyst is triggered and escorted by electrochemical reaction occurred in the SOFC. This breeding phenomenon can enhance cell performance obviously. The breeding reactions are described as, 1) $NiO + H_2 \rightarrow Ni + H_2O$, and 2) $Ni + O^{2-} \rightarrow NiO + 2e^-$. Hence, Ni can be a reactant to generate currents following the above reactions. The cell performance change and microstructure analysis allows us to verify functions of nano-scale catalyst at anode during the cell operation. The functions and breeding cycle of nano-scale nickel catalyst at anode are shown in Fig. 10[29]. This breeding of nano-scale catalyst shows that the issue of catalyst sintering and agglomeration is different between catalytic reactions of purely chemical and electro-chemical systems. The aged anode catalyst can be rejuvenated by the application of a strong reducing environment on the anode side or periodic reduction treatment of the anode side, such as the electrochemical reduction includes applying an external voltage to the cell in a reverse current direction at anode or a chemical reduction by periodically idling the cell in the dry hydrogen gas, using the schemes of operation adjustment to enhance the cell durability. It is consistent with the scheme proposed by Hickey *et al.*[30], B). Establishment of the measurement process and equipment for determination of the optimum contact pressure among stack units of a high performance solid oxide fuel cell stack (SOFC-stack) [31]. A measuring process and apparatus for determination of the optimum contact pressure among components of a solid oxide fuel cell stack in the packaging process is developed, as shown in Figs. 11-13. The method may provide the best value of contact pressure employed for the fabrication of cell stack to achieve the maximum output powder density without damaging the SOFC-MEA.

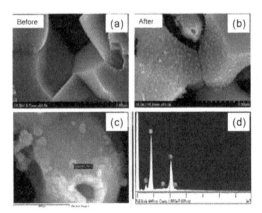

Fig-10 Breeding phenomena of bred nano-scale nickels at anode by electrochemical reaction: (a) before cell operation, (b) after cell operation, (c) & (d) EDS analysis of the bred Ni particles [29].

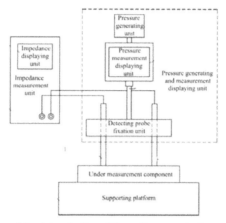

Fig. 11 Equipment for determination of the optimum contact pressure among stack units of SOFC

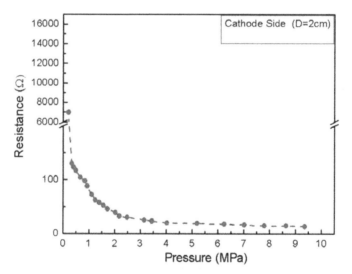

Fig. 12 Contact pressure vs. cathode resistance of SOFC- MEA

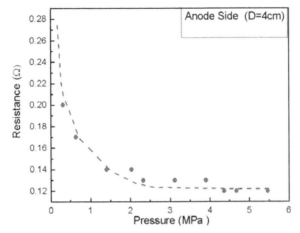

Fig. 13 Contact pressure vs. anode resistance of SOFC- MEA

2. Application of SOFC-MEA Technology

Key technologies of SOFC-MEA are essentially described as: 1) fine ceramic powder preparation, 2) fabrication of thick and thin film of ceramic layers with properties to match the requirements of component of products, 3) thin film (nano/micron-scale) of ceramic layer with characteristics of gas tight used at high temperature (near 1000 °C), and 4) Lamination and sintering technologies for fabrication of the multiple layers of composite. In addition to the application of SOFC-MEA for combined heat and power (CHP) generation , energy conversion and storage via solid oxide electrolysis cells (SOEC)[32-35], these technologies can be applied for development of the essential equipments or energy conversion and storage systems to solve the related problems of energy and greenhouse effect. Key applications of SOFC-MEA technology at INER are described as follows.

2-1. Research and development of the unit equipment of oxygen transport membrane reactor
    (OTMR) and hydrogen transport membrane reactor (HTMR)

INER executes the project of the IGCC technology for biomass and coal based central power generation and large demonstration systems as well as the carbon dioxide capture and storage (CCS) [1]. CCS technologies attract a lot of attention because they would reduce our CO_2 emissions to the atmosphere while continuing to use fossil fuels. Three processes are available for the capture of CO_2 from large point sources: Post-combustion capture, pre-combustion capture and oxygen-fired combustion[2]. In post- combustion systems, CO_2 is separated from the flue gases produced by the combustion of the primary fuel in air. In pre-combustion systems, the primary fuel is treated in a reactor with steam and air or oxygen to produce synthesis gas (CO and H_2). Oxygen-fired combustion proceeds by an approach where air separation precedes combustion. The hydrocarbon fuel is then combusted in a mixture of O_2 and CO_2 rather than air to produce an exhaust of CO_2 and water vapour because oxygen is used instead of air, the nitrogen part and its combustion products are ideally eliminated from the exhaust gas stream. Hence, the OTMR technology is critical to improve the process by supply of pure oxygen from air separation unit (ASU) and solve the problems. Gallucci1 et al.[36] investigated the main aspects related to the patents on hydrogen production through membrane reactors and reported that the membrane reactors are a modern configuration which integrates reaction and separation units in one vessel and results in a tremendous degree of process intensification[37]. A membrane reactor, such as SOFC, represents a modern configuration in which an integrated reaction/separation unit has many potential advantages for reduction of costs. Sander et al.[38] proposed a new concept of an IGCC process with an integrated oxygen transport membrane reactor and CO_2 capture and showed that the operating conditions of the membrane reactor affect the overall performance of the IGGC process tremendously. The roles of HTMR and OTMR are critical for the developments of IGCC and CCS systems. It is obvious that the related technologies of SOFC-MEA are critical techniques for developments of HTMR and OTMR applied at the IGCC and CCS systems for carbon purification program and   enhancement of the technologies of green energy.

2-2. Development of total solution for energy conversion and storage

The conflict between the increases of energy request for human civilization and CO_2 emission to trig the greenhouse effect must be solved. Three strategies are proposed and well recognized : 1) increasing the energy efficiency or a change in the primary energy source to decrease the amount of CO_2 emitted., 2) the use of CO_2 as a chemical feedstock in different applications-- transform it and use it, 3) the development of new technologies for capture and sequestration of CO_2 for storage[2]. The utilization of CO_2 as a raw material to reduce the atmospheric CO_2 loading in conjunction with the CCS technology is the critical method in this regard. Essentially, this idea is consistent with our scope. A process and apparatus of "Solid Oxide Fuel Cell-CO_2 Energy Conversion Cycle (referred to as SOFC-CO_2-ECC)" are invented to adopt CO_2 as energy sources from waste/stock gas or convert and fix it in the useful compounds at INER[39-42]. The processes of SOFC-CO_2-ECC and experimental results are shown in Fig 14 and Figs 15/16, respectively. Due to simultaneously high-temperature (700-1000°C) catalytic and electrochemical reactions in the SOFC, CO_2, a compound having extreme chemical stability, is cracked at about 800 °C following a chemical reaction below: $CO_2 \rightarrow CO_{(g)} + 1/2O_2$, to generate $CO_{(g)}$ and O_2, then O_2 electrochemically reacts with H_2 (or other hydrocarbons, such as methane) in the SOFC for power generation. As such, CO_2 may serve as a power source material in an overall reaction of the SOFC, and converted into a very useful energy source or compound CO, which can be derived into useful compounds or energy source, for example, aldehydes and alcohols, for regeneration of energy source, so as to complete an overall energy conversion carbon cycle of CO_2 → CO → derivatives of CO (fixation of CO_2)→ generation of energy source → CO_2, and achieve zero emission of CO_2. Technical solutions, auxiliary materials and equipments of this invention are capable of effectively solving the problem of global extinction of organisms caused by greenhouse effect of CO_2 and grasping the "Right of Carbon Emission Trading" issues. More efforts must be input to solve the related problem, such as the coking reaction to degrade the SOFC-MEA/stack. One test system is operating for efficiency/durability identification at this stage.

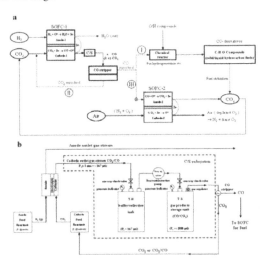

Fig-14 The principle and process flow structure of the SOFC-CO2-ECC system (a).
Collection and storage subsystem for gas products of the cathode outlet stream (b).

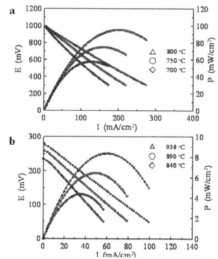

Fig. 15 The cell performance test results of SOFC-1 for CO$_2$ cracking reaction.
Quality test of fuel cell via H$_2$ fuel and O$_2$ oxidant of the air (a).
Performance results of fuel cell for CO$_2$ cracking reaction (b).

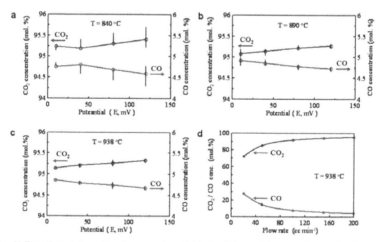

Fig. 16 Experimental results of conversion yields for CO$_2$ cracking reaction under diversely
operation conditions on SOFC-1. Concentrations of CO$_2$ and CO versus potential at
840 °C (a), 890 °C (b), and 938 °C (c), respectively. The flow rates of reactants of N$_2$+
H$_2$ (anode) and CO$_2$ (cathode) are 200 ml min^{-1}. Concentrations of CO$_2$ and CO versus
reactant flow rates at 938 °C with potential of 10 mV (d).

RESULTS AND DISCUSSION

Substantial progresses are made at INER in MEA, stack, and power system developments. Innovative inventions of the intellectual property/patents have been granted from Taiwan, Japan, the USA and European Union. Establishments of strong capability and firmed facilities both hardware and software will facilitate the industrial development on SOFC technology. Nevertheless, continuous improvements are required to make the SOFC become commercially viable with high reliability, longevity, and low cost. The conclusions are described as follows:

(1) The application of SOFC technology to increase the energy conversion efficiency of the hydrocarbon fuels from chemical energy to electrical power, to benefit the target of carbon reduction or to mitigate the greenhouse effect, is well confirmed. The extended application of SOFC-MEA technology in the issue of membrane reactors, such as HTMR and OTMR, is critical for enhancement of the project progresses of IGCC and CCS. A process and apparatus of SOFC- CO_2-ECC has been proposed and tested at INER to achieve the efficacy of environmental protection via solving the problem of CO_2 greenhouse effect. More efforts must be input to upgrade the system efficiency and integration of the pilot system.

(2) It is a critical issue to have a SOFC power system with natural gas as the fuel. Essentially, we are on the way to commercialization. Both quality and quantity of MEAs/Stacks are the bases of the development of SOFC technology. One small-scale production line of INER's SOFC-MEA is operated in 2012 with yield of near 100%. Meanwhile, reliability and power performance of MEAs are continuously improved further to meet commercial target, i.e., degradation rate less than 0.1%/khr and operating lifetime higher than 40,000 hours.

(3) Durability and reliability of the SOFC-MEA/stack/power generation are critical prerequisites for fabrication of a robust SOFC system. The approach from laboratory to demonstrative pilot and then to commercialization is essential to move forward the SOFC technology. Meanwhile, the government has to provide new initiatives, such as: deregulations on power net, power purchase agreements, subsides and incentives etc., to promote the high efficiency energy technologies. Now it is the right time to have the national/global cooperation on the projects to organize industrial partners for the commercialization of the SOFC technology.

ACKNOWLEDGEMENT

Financial support from the National Energy Program of Taiwan Government is gratefully acknowledged.

REFERENCE

[1]R.Y. Lee, Y.N. Cheng, C.S. Hwang, and M.C. Lee, Development of SOFC Technology at INER, *Institute of Nuclear Energy Research*, 10th European SOFC Forum 2012, Lucerne Switzerland, 2012.

[2]M. Mikkelsen, M. Jørgensen, and F. C. Krebs, The teraton challenge: A Review of Fxation and Transformation of Carbon Dioxide, *Energy Environ. Sci.*, **3**, 43–81 (2010).

[3]R.J. Yang, M.C. Lee, J.C. Chang, T.N. Lin, Y.C. Chang, W.X. Kao, L.S. Lee, and S.W. Cheng, Fabrication and characterization of a $Sm_{0.2}Ce_{0.8}O_{1.9}$ electrolyte film by the spin-coating method for low-temperature anode-supported solid oxide fuel cells, *Journal of Power sources*, **206**, 111–118 (2012).

[4]J. Sunarsoa, S. Baumannb, J.M. Serrac,W.A. Meulenbergb, S. Liua, Y.S. Lind, and J.C. Diniz da Costa a , Mixed Ionic–Electronic Conducting (MIEC) Ceramic-Based Membranes for Oxygen Separation, *Journal of Membrane Science*, **320**, 13–41 (2008).

[5]V.V. Kharton, F.M.B. Marques, Mixed ionic–electronic conductors: Effects of Ceramic Microstructure on Transport Properties, *Curr. Opin., Solid State Mater. Sci.*, **6**, 261-269 (2002)

[6]Y.C. Chang, M.C. Lee, T.N. Lin, W.X. Kao, Preparation of Nano-scale/SOFC Grade Yttria-stabilized Zirconia (YSZ) Material: A Quasi-Optimization of The Hydrothermal Coprecipitation Process, *Int. J. of Appl. Ceramic Tech.*, **5**, 557-567 (2008).

[7]J.C. Chang, M.C. Lee, R.J. Yang, Y.C. Chang, T.N. Lin, C.H. Wang, W.X. Kao, and L.S. Lee, Fabrication and Characterization of SDC-SSC Composite Cathode for Anode Supported Solid Oxide Fuel Cell, *J. Power Sources*, **196**, 3129-3133 (2011).

[8]W.X. Kao, M.C. Lee, T.N. Lin, C.H. Wang, and Y.C. Chang, Fabrication and Characterization of a $Ba_{0.5}Sr_{0.5}Co_{0.8}Fe_{0.2}O_{3-x}$-gadolinia-doped Ceria Cathode for an Anode-supported Solid-oxide Fuel Cell, *J. Power Sources*, **195**, 2220-2223 (2010).

[9]C.H. Wang, M.C. Lee, T.N. Lin, W.X. Kao, and Y.C. Chang, Process To Produce Fine Ceramic Powder through A Chemical Reactor With Powder Collection Device, US patent No.: 8,287,813 B2 (Oct.16,2012).

[10]L.S. Lee, M.C. Lee, T.N. Lin, W.X. Kao, J.C. Chang, R.J. Yang, and S.W. Cheng, Fabrication and Performance Test of an Anode-supported Solid Oxide Fuel Cell with Al_2O_3 Additive in Anode, 2012 MRS-T ANNUAL MEETING, 0015-0017 (2012).

[11]S.H. Nien, C.S. Hsu, C.L. Chang, and B.H. Hwang, Preparation of $BaZr_{0.1}Ce_{0.7}Y_{0.2}O_{3-x}$ Based Solid Oxide Fuel Cells with Anode Functional Layers by Tape Casting, *FUEL CELLS*, **11**, 178-183 (2011).

[12]T. Norby, Solid-state protonic conductors: principles, properties, progress and Prospects, *Solid State Ionics*, 125, 1-11 (1999).

[13]S. Yamaguchia, K. Nakamuraa, T. Higuchib, S. Shinc, and Y. Iguchia, Basicity and Hydroxyl Capacity of Proton-conducting Perovskites, *Solid State Ionics*, **136–137**, 191–195 (2000).

[14]C.S. Hwang, C.H. Tsai, J.F. Yu, C.L. Chang, J.M. Lin, Y.H. Shiu, and S.W. Cheng, High Performance Metal-supported Intermediate Temperature Solid Oxide Fuel Cells Fabricated by Atmospheric Plasma Spraying, *Journal of power Sources*, **196**, 1932–1939 (2011).

[15]D. Nguyen, R. Oswal, T. Armstrong, E. E. Batawi, Heterogeneous Ceramic Composite SOFC Electrolyte", Assignee: Bloom Energy Corporation, US Patent, Pub. No: US 2008/0261099 A1, Pub. Date: Oct. 23, 2008.

[16]R.J. Yang, M.C. Lee, J.C. Chang, T.N. Lin, W.X. Kao, and L.S. Lee, A Brief Review of Scandia-stabilized Zirconia as Electrolytes for SOFCs, INER-9470R, INER, Taiwan, September 2012.

[17]L.C. Wen, H.Y. Hsieh, Y.H. Lee, S.C. Chang, H.C.I. Kao, H.S. Sheu, I.N. Lin, J.C. Chang, M.C. Lee, and Y.S. Lee, Preparation of Compact Li-doped $Y_2Ti_2O_7$ Solid Electrolyte, *Solid State Ionics*, **206**, (2012)

[18]C.H. Wang, M.C. Lee, T.N. Lin, Y.C. Chang, W.X. Kao, and L.F Lin, The Innovation Process for Anode Treatment of Solid Oxide Fuel Cell – Membrane Electrode Assembly (SOFC-MEA) to Upgrade the Power Density in Performance Test, US patent No.: 7,815 843 B2 (Oct.19.2010)/ ROC Patent No. I 32696969 (01, 02, 2011) /European Patent NO : EP2107630A1 (28-September-2011)/ (Japan patent-Pending).

[19]Y.C. Chang, M.C. Lee, W.X. Kao, C.H. Wang, T.N. Lin, J.C. Chang, and R.J. Yang, Characterizations of The Anode-supported Solid-oxide Fuel Cells With an Yttria Stabilized Zirconia Thin Film by The Diagnosis of The Electrochemical Impedance Spectroscopy, *J. Electrochemical Soc.*, **158(3)**, B259-B265 (2011).

[20]W.X. Kao, M.C. Lee, Y.C. Chang, T.N. Lin, C.H. Wang, and J.C. Chang, Fabrication and Evaluation of The Electrochemical Performance of The Anode-supported Solid Oxide Fuel Cell With The Composite Cathode of $La_{0.8}Sr_{0.2}MnO_{3-\delta}$–Gadolinia-doped Ceria oxide/$La_{0.8}Sr_{0.2}MnO_{3-\delta}$, *Journal of Power Sources*, **195**, 6468–6472 (2010).

[21]M.C. Lee, W.X. Kao, T.N. Lin, Y.C. Chang, and C.H. Wang, A Novel Synergistic Process and Recipe for Fabrication of a High Integrity Membrane Electrode Assembly of Solid Oxide Fuel Cell., European patent No.: EP 2045858 A1 (07.04.2010)/ ROC Patent No.: I 326933 (01,07,2010)./ US patent No.:7914636 (2011) /Japan Patent No.: 特許第5009892號 (Oct..05.2012)

[22]M.C. Lee, C.H. Wang, W.X. Kao, Y.C. Chang, T.N. Lin, and Y.H. Shiu, Fabrication and Performance Test for Durability Evaluation of INER-SOFC-MEA(W-12) via Spin Coating Process, INER-5353R, INER, Taiwan, July 2008.

[23]T.N. Lin, M.C. Lee, L.S. Lee, W.X. Kao, J.C. Chang, and R.J. Yang, Fabrication and Performance Test of an $Y_{0.2}Zr_{0.8}O_{3-\delta}$ Electrolyte Film by The Spin-coating Method for an Intermediate-temperature Anode-supported Solid Oxide Fuel Cells, Submitted, (2012).

[24]W.X. Kao, M.C. Lee, T.N. Lin, C.H. Wang, Y.C. Chang, and L.F Lin, A Novel Process for Fabrication of a Fully Dense Electrolyte Layer Embedded in a High Performance Membrane Electrode Assembly (MEA) (Unit Cell) of Solid Oxide Fuel Cell, EURO-Patent NO:EP 2083465A1 (101-08-01).

[25]Y.C. Chang, M.C. Lee, C.H. Wang, W.X. Kao, T.N. Lin, and L.F Lin, Formulation of Nano-scale Electrolyte Suspensions and Its Application Process for Fabrication of Solid Oxide Fuel Cell-membrane Electrode Assembly (SOFC-MEA), US patent No.: 8,158,304 B2 (April 17,2012)

[26]M.C. Lee, T.N. Lin, W.X. Kao, R.J. Yang, and L.S. Lee, Performance and Durability Evaluation of the Anode-supported Solid Oxide Fuel Cell after Long-term operation, Paper number: X00-002, Proceedings of 2012 Taiwan Symposium on Carbon Dioxide Capture, Storage and Utilization, November 25~27, Taipei.

[27]T.N. Lin, M.C. Lee, R.J. Yang, J.C. Chang, W.X. Kao, and L.S. Lee, Chemical State Identification of Ce^{3+}/Ce^{4+} in The $Sm_{0.2}Ce_{0.8}O_{2-\delta}$ Electrolyte for an Anode-supported Solid oxide Fuel Cell After Long-term Operation, *Materials Letters*, **81**, 185–188 (2012).

[28]E.D. Wachsman, K.T. Lee, Lowering the Temperature of Solid Oxide Fuel Cells, *Science*, **334**, 935-939 (2011).

[29]C.H. Wang, M.C. Lee, T.J. Huang, Y.C. Chang, W.X. Kao, and T.N. Lin, Breeding Phenomenon of Nickel in Anode of Solid Oxide Fuel Cell via Electrochemical Reaction, *Electrochemistry Communications*, **11**, 1381-1384 (2009).

[30]D. Hickey, C. Karuppaiah, and J. McElroy, Reduction of SOFC Anodes To Extend Stack Lifetime", US Patent No: US7, 514,166 B2, (Apr. 7, 2009).

[31]M.C. Lee, W. X. Kao, T.N. Lin, S.H Wu, Y.C. Chang, R.Y. Lee, and C.H. Wang, A Novel Measurement Process and Equipments for Determination of the optimum Contact Pressure Among Stack Units of a High Performance Solid Oxide Fuel Cell Stack (SOFC-Stack)，Pending patent (1) US: 12/874307(2010-09-02) (2) EU: 10175-75-277.2. (2010-09-03))

[32]K. Eguchi, T. Hatagishi, and H. Arai, Power Generation and Steam Electrolysis Characteristics of an Electrochemical Cell with a Zirconia- or Ceria-based Electrolyte, *Solid State Ionics*, **86-88**, 1245-1249 (1996).

[33]Meng Ni, Michael K.H. Leung, Dennis, and Y.C. Leung, Technological Development of Hydrogen Production by Solid Oxide Electrolyzer Cell (SOEC), *International Journal of Hydrogen Energy*, **33**, 2337 – 2354 (2008).

[34]P. Ioraa, P. Chiesab , High Efficiency Process for The Production of Pure Oxygen Based on Solid Oxide Fuel Cell–solid Oxide Electrolyte Technology, *Journal of Power Sources*, **190** (2009) 408–416.

[35]J. Udagawa, P. Aguiar, and N.P. Brandon, Hydrogen Production Through Steam Electrolysis: Model-based Steady State Performance of a Cathode-supported Intermediate Temperature Solid Oxide Electrolysis Cell, *Journal of power Sources*, **166**, 127-136 (2007).

[36]F. Galluccil, A. Basile, A. Iulianelli ,and H.J.A.M. Kuipersl, A Review on Patents for Hydrogen Production Using Membrane Reactors, *Recent Patents on Chemical Engineering*, **2**, 207-222 (2009).

[37]M. Czyperek, P. Zapp, H. J. M. Bouwmeester, M. Modigell, K. Ebert, I. Voigt, W.A. Meulenberg, L. Singheiser, D. Stöver, Gas separation membranes for zero-emission fossil power plants: MEM-BRAIN, *J. Membr. Sci.*, 359, 149−159 (2010).

[38]F. Sander, R. Span, Modelling of an Oxygen Transport Membrane for an IGCC Process with CO_2 Capture, Proceedings GHGT8, Trondheim, 2006.

[39]C.H. Wang, T.J. Huang, M.C. Lee, W.X. Kao, Y.C. Chang, T.N. Lin, J.C. Chang, and R.J. Yang, Direct Methane Oxidation and Electrochemical Reduction of Carbon Dioxide with Power Generation via Solid Oxide Fuel Cell, 28th Taiwan Symposium of Catalysis and Reaction Engineering, Taipei, Taiwan, June 2010.

[40]T.J. Huang, C.L. Chou, Electrochemical CO_2 Reduction With Power Generation in SOFCs With Cu-added LSCF–GDC Cathode, *Electrochemistry Communications,* **11**, 1464-1467 (2009).

[41]M. Homel, T.M. Gur, J.H. Koh, A.V. Virkar, Carbon Monoxide-fueled Solid Oxide Fuel Cell , *J. of Power Sources*, **195** 6367– 6372 (2010).

[42]M.C. Lee, C.H. Wang, Y.C. Chang, W.X. Kao, T.N. Lin, J.C. Chang, R.J. Yang, and L.S. Lee, Process and Apparatus of CO_2 Energy Source Adopted in Solid Oxide Fuel Cell - CO_2 Energy Conversion Cycle", Pending Patent, Application No: (1) US 12/973,507 / Pub. No. 2012 / 0115067 A1 (Pub. Date: May 10, 2012).

AQUEOUS PROCESSING ROUTES FOR NEW SOFC MATERIALS

Maarten C. Verbraeken, Mark Cassidy, John T.S. Irvine
University of St Andrews, School of Chemistry, North Haugh, St Andrews, Fife, KY16 9ST, United Kingdom

ABSTRACT
This paper discusses the development of new polar inks for screen printing onto green tapes, which have been produced by aqueous methods. Conventional organic inks have shown incompatibility with these aqueous tapes, due to de-wetting. Here it is shown that moving to a more polar system, improves the wetting characteristics of the prints.

The new screen printing inks are based on glycerol-water and propylene glycol-water based solvent systems, with PVP (polyvinyl pyrrolidone) as binder. Rheological data is presented for both polar ink vehicles, as well as inks and comparisons are made with a typical organics based vehicle. A large degree of pseudoplasticity was found in all polar inks. Some thixotropy is also found in binary glycerol-water based inks as well as the polypropylene glycol based inks. This thixotropy is believed to be beneficial in obtaining uniformly thick prints. More studies are required to understand the interaction between solvents and binder, as this seems key in the different levels of thixotropy observed. Optimising the rheological behaviour and in particular the degree of thixotropy in these inks should lead to further improvements in printing quality. Currently, best printing results are obtained for inks based on propylene glycol with PVP binder.

INTRODUCTION
Conventional state-of-the-art Ni cermet anodes for Solid Oxide Fuel Cells (SOFC) perform well in hydrogen and high steam reformate fuels, but still suffer a number of drawbacks. Their poor redox stability, tendency for coking in hydrocarbon fuels and low sulphur tolerance has led many researchers to look for alternative anode materials. One promising approach is to replace the cermet with an electronically conducting ceramic to provide both structural support and current collection, whereas electrocatalytic activity can be obtained through impregnation of metal solutions into the porous scaffold. A-site deficient strontium titanates show high n-type conductivity and therefore offer promise as the electronically conducting backbone. Previous work in our research group has shown promising results for a La and Ca co-doped $SrTiO_3$ anode in an electrolyte supported cell (ESC) configuration, into which nickel and ceria had been impregnated to serve as fuel oxidation catalysts. Fuel cell performances in hydrogen were similar to those for Ni-cermet anodes, but with much improved redox stability[1].

In this work we are now applying the same approach to an anode supported cell (ASC) configuration. The A-site deficient $La_{0.20}Sr_{0.25}Ca_{0.45}TiO_3$ ($LSCT_{A-}$) shows good mechanical strength and should therefore be suitable as substrate material. To this end, an aqueous tape casting process was developed initially, as an environmentally friendly alternative to the organics based process. However, the aqueous processing also allows for using a wider range of polymeric pore formers that would otherwise dissolve in the solvent. This in turn allows for a more precise control of the porous scaffold's microstructure for optimal electrode performance. The next stage of this work is to screen print electrolyte layers onto green aqueous tapes and subsequently co-sinter. Screen printing inks based on conventional organic vehicles have proven to be unsuitable for printing onto green aqueous tapes due to de-wetting problems. This paper will discuss the development of new inks based on a more polar solvent system to improve compatibility. In particular, screen printing ink vehicles based on glycerol/propylene glycol – water mixtures with polyvinyl pyrrolidone as binder were studied, as their rheological behaviour showed similarities to typical conventional ink vehicles based on solvent-binder systems, such as terpineol with polyvinyl butyral[2].

EXPERIMENTAL
Powders
    The A-site deficient perovskite material $La_{0.20}Sr_{0.25}Ca_{0.45}TiO_3$ ($LSCT_{A-}$) was synthesised by Topsoe Fuel Cells, using the drip pyrolysis method[3]. The resulting powder is a nanosized, high surface area powder (40 m^2/g). The powder is calcined prior to ceramic processing, resulting in a d_{50} particle size ranging between 0.7 and 2.0 μm. 8-YSZ ($Zr_{0.85}Y_{0.15}O_{1.92}$) was purchased from Pi-Kem Ltd. and has a d_{50} particle size of 0.2 μm.

Aqueous Tape Casting
    Both Duramax D3005 (Rohm & Haas) and Hypermer KD6 (Uniqema) have been employed as dispersants with deionised water as the solvent. Polyethylene glycol (PEG, av. M_w 300, Sigma Aldrich) and glycerol (Alfa Aesar) are used as plasticisers with a high molecular weight PVA (polyvinyl alcohol, Alfa Aesar) as the binder. A molecular defoamer (WT001, Polymer Innovations) is also added to prevent extensive foaming and to help in releasing the tapes from the mylar carrier. The tape casting slips are produced in a three-step procedure. First the ceramic powder is dispersed in deionised water by rotary ball milling, using either dispersant for 24 hours. After this step the organics are mixed in by ball milling at a reduced speed to prevent degradation of the binder polymer. The PVA is pre-dissolved in deionised water to make a 15 wt% solution. Finally the slip is de-aired by either slow rolling or leaving it under a mild vacuum. Viscosities of the slips were controlled by adjusting the organics and solids content and optimised for the casting process. This aqueous process route proved to be flexible and could be used to produce both dense and porous substrates. Tapes were cast using a bench scale tape caster from Mistler and Co. onto an uncoated mylar carrier film. Drying of the tapes was in static air in the absence of any form of heating.

Polar screen printing ink development
    The ceramic powder used in all inks was the above mentioned 8-YSZ. The 8-YSZ was first dispersed in isopropyl alcohol (IPA) with small amounts of PVP K30 as dispersant, using rotary ball milling. After this dispersion step, the various vehicle combinations under study were added. This mixture was continuously agitated until the IPA had evaporated to yield the final ink. The ceramic loading in all inks was 69 wt%. The ink vehicles studied here were based on glycerol – water and propylene glycol (Alfa Aesar) – water mixtures with various amounts of polyvinyl pyrrolidone (PVP K90, $M_w \sim 1,400,000$, Alfa Aesar) as a binder. Rheological tests were also carried out on conventional organic vehicles, consisting of terpineol (Sigma Aldrich) and PVB (polyvinyl butyral, Butvar®, Sigma Aldrich). Rheological behaviour of ink vehicles and inks was measured on a Brookfield DVIII Ultra Rheometer using a small sample adapter. All viscosity data was recorded at room temperature. Printing was carried out onto green $LSCT_{A-}$ substrates, using a DEK248 semi automatic printer.

    Tapes and printed layers are co-fired using a non-optimised two-step sintering profile as similarly reported by Wright and Yeomans[4]. Chen and Wang were the first to propose two-step sintering as a means to achieve high density ceramics without grain growth[5]. Using a single step firing protocol, leads to dense $LSCT_{A-}$ substrates. The two-step sintering profile could provide both porous anode supports as well as dense electrolytes. The co-sintered ceramics and printed layers were studied by SEM (Jeol JSM 6700F FEG-SEM).

RESULTS
Aqueous Tape Casting

Both Duramax D3005, a polyelectrolyte with electro-steric dispersing functionality, as well as Hypermer KD6 (steric stabilisation) were found effective in stabilising 8-YSZ or $LSCT_A$ in deionised water. However, we noticed an interaction of D3005 with the remaining organics during the second stage of the slip production, resulting in some minor agglomeration of the ceramic powder in the final tape. These defects are likely to cause a weakening of the mechanical strength after sintering of the green tapes and are therefore not desirable. Since KD6 did not show this negative effect, it was used in the production of all aqueous slips. Figure 1 shows a cross section of a $LSCT_A$ dense ceramic obtained by our aqueous method. Some minor defects are still present, which are believed to be due to residual small air entrapments in the slips. Overall however, the combination of the various organics in this formulation results in homogeneous tapes with good laminability for further processing.

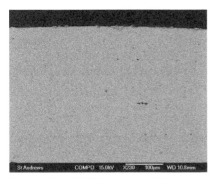

Figure 1: Dense LSCT substrate made by aqueous tape casting. Minor defects probably due to entrapped air in green tapes

Polar screen printing inks

The viscosities of the two solvent systems, i.e. glycerol – water and propylene glycol – water were first measured and the results are shown in Figure 2 and Figure 3. The binary mixtures behave as Newtonian fluids. The results are in good agreement with previously published data by Shankar and Kumar[6], and Sun and Teja[7]. Shankar and Kumar already described the highly non-linear dependence of viscosity with weight fraction x_g in the glycerol – water system. This is commonly observed in polar mixtures and is due to the strong interactions of the two molecules. A non-linear dependence is also observed in the propylene glycol – water binary mixture, but on a much smaller scale.

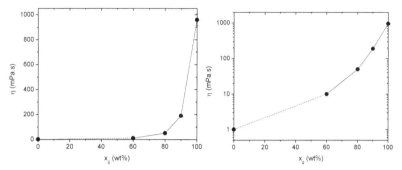

Figure 2: Variation of dynamic viscosity with glycerol content (x_g) in glycerol – water mixture. Data measured at 22°C

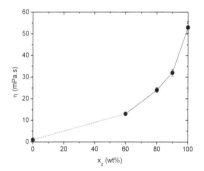

Figure 3: Variation of dynamic viscosity with propylene glycol content (x_{pg}) in propylene glycol – water mixture. Data measured at 22°C

Upon adding PVP K90 as polymeric binder, the rheological behaviour changes from Newtonian to slightly pseudoplastic. As expected, the pseudoplasticity increases with increasing binder content, due to the effect of alignment of polymer molecules along the shear direction. The pseudoplasticity is most pronounced in vehicles with low water content. Figure 4 show some typical curves for both the glycerol – water and propylene glycol – water mixtures with 5 – 10 wt% PVP K90.

The viscosity dependence with varying binder content is shown in Figure 5 for both binary solvent mixtures. The same dependence is shown for the terpineol – PVB system.

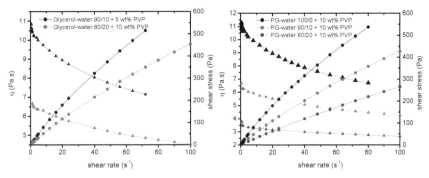

Figure 4: Rheological behaviour of glycerol-water (left) and propylene glycol-water (right) with 5 – 10 wt% PVP K90 binder. Data recorded at 20°C.

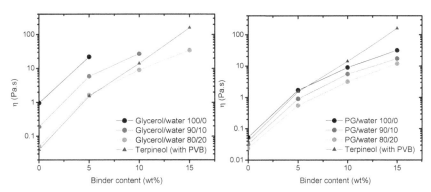

Figure 5: Dependence of viscosity in binary solvent vehicles, with binder content (PVP K90). Data taken at constant shear rate of 20 s⁻¹.

Four vehicles were finally selected to form the basis of screen printing inks and are listed in Table 1. All inks have a solids loading of 69 wt% (8-YSZ). Three inks contain 5 wt% PVP K90, whereas the ink vehicle using 100% glycerol as solvent (i.e. no water) contains only 2 wt% binder, due to limited solubility. Figure 6 and Figure 7 show the rheological behaviour of the four inks. It can be seen that all inks show a large degree of pseudoplasticity. In all cases the viscosity drops by more than an order of magnitude when the shear rate increases from 0.2 s⁻¹ to 20 s⁻¹. Interestingly, ink #4 has slightly higher viscosity than ink #3, which seems contradictory to the results presented in Figure 5.

Furthermore, all inks show some degree of thixotropy too, apart from the ink based on glycerol with 2 wt% PVP. Thixotropy is evidenced by the hysteresis in the shear stress vs. shear rate curves (Figure 7). This hysteresis is absent in ink #1. This thixotropic effect was further studied by measuring the viscosity in the time domain at constant shear rate. As expected, ink #1 shows very little time dependency, with a drop in viscosity of only ~5% over the course of 25 seconds, at a low shear rate of 0.8 s⁻¹. At the higher shear rate of 2.4 s⁻¹, the viscosity is constant within 1%. Of the remaining inks, the water containing ones show most variation in viscosity with time. At 0.8 s⁻¹, ink#2 and ink #4 show a decrease in viscosity of 14% and 16%, respectively, over a 60 s time interval, whereas ink #3 shows an 11% decrease over the same time interval. Little change in viscosity is recorded beyond 60s. However, at higher shear rate $(2.4\ s^{-1})$ only small changes in viscosity can be observed with time for inks #2-4; a decrease of 2 – 3% is recorded over less than 30 s, after which a steady state is observed. The time dependence of viscosity in the various inks is also shown in Figure 8. All curves seem to follow a general logarithmic trend according to $\eta = -A \cdot \ln(t) + B$, where A and B are constants. The magnitude of A is measure of the degree of thixotropy. Overall, the thixotropy in all inks seems fairly limited.

Table 1: Overview of four selected 8-YSZ inks with polar vehicles

	Ink #1	Ink #2	Ink #3	Ink #4
Solids loading (wt%)	69	69	69	69
Vehicle solvent (w/w)	Glycerol (100/0)	Glycerol/water (80/20)	Propylene glycol (100/0)	Propylene glycol/water (90/10)
Binder content in vehicle (wt%)	2	5	5	5
A (@ 0.8 s^{-1}, fit)	2.1	6.0	4.6	7.2
A (@ 2.4 s^{-1}, fit)	0.3	1.8	1.6	1.3

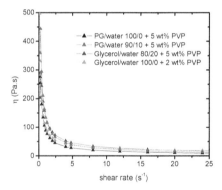

Figure 6: Viscosity vs. shear rate for inks #1-4. All inks show large degree of pseudo plasticity.

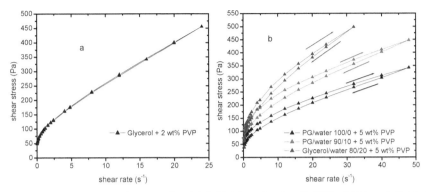

Figure 7: Shear stress vs. shear rate for a) ink #1 (glycerol + 2 wt% PVP vehicle), and b) ink #2-4 (glycerol/water 80/20 + 5 wt% PVP vehicle, propylene glycol/water + 5 wt% PVP vehicles). Inks #2-4 show hysteresis, which is indicative of thixotropy. Ink #1 does not exhibit thixotropic behaviour

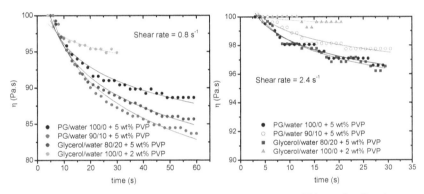

Figure 8: Time dependence of viscosity at two different shear rates. Thixotropic effect decreases with increasing shear rate

All inks have been printed onto green LSCT$_{A-}$ substrates. No de-wetting occurs for these polar inks. After firing, the printed layers were studied using SEM. The micrographs can be seen in Figure 9. Figure 9a shows the layer printed using ink #1. It shows a non-uniform thickness, with peaks and troughs. The printed layer produced by ink #2, i.e. a binary glycerol/water based ink, is slightly improved as compared to ink #1, with a less pronounced peak/trough profile, as shown in Figure 9b. There is still a degree of non uniformity in the thickness however. The layers obtained by using either propylene glycol or a binary propylene glycol/water based ink show a uniform thickness, as shown in Figure 9c and d. There is currently variable porosity in the electrolyte layers as well as the anode supports. This will hopefully this be remedied by fine tuning the firing protocol as well as optimising the LSCT$_{A-}$ powder characteristics (i.e. particle size etc.). The characteristics of the printed layers as observed in Figure 9 are solely due to use of different inks.

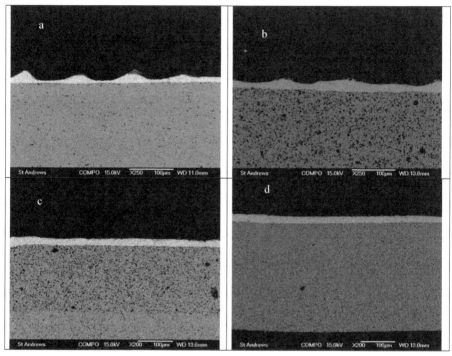

Figure 9: a) Print with ink #1 (comprising glycerol + 2 wt% PVP vehicle); b) Print with ink #2 (glycerol/water 80/20 +5 wt% PVP); Print with ink #3 (propylene glycol + 5 wt% PVP); d) Print with ink #4 (propylene glycol/water 90/10 + 5 wt% PVP. The distance between peaks in a) match the screen mesh spacing

DISCUSSION
Aqueous tape casting
    Our aqueous tape casting method has produced both dense and porous flat ceramics, with only minor defects, due to air entrapment in the slips. Fine tuning of the organics content, combined with better de-airing techniques (e.g. mild vacuum treatment) should remedy this problem. Currently, the maximum, homogeneous, thickness of single cast tapes attainable is limited to ~250 – 300 μm. Edge effects become apparent when trying to increase the thickness further, resulting in inhomogeneity. Further optimisation is required here to obtain single cast anode supports.

Polar screen printing
    First tests with polar screen printing inks were performed using the single solvent vehicle, ink #1, i.e. glycerol with 2 wt% PVP K90 binder. This ink caused large non-uniformity in the thickness of the prints. The spacing between peaks matches well with the screen mesh spacing. The non-uniformity can be related to a combination of high degree of pseudo plasticity and the lack of thixotropy in this ink as observed in Figure 7. Thixotropic systems retain some of the energy when the applied shear is removed, leading to a temporary reduction of viscosity, which takes time to recover to its initial value. This in turn allows for smoothing and levelling of the

printed layers after passing through the screen. When this thixotropic behaviour is not present the ink rapidly returns to the initial high viscosity state in the absence of a shear force, preventing flow and therefore levelling. A secondary drawback of using this glycerol based ink, was the poor drying characteristics, caused by glycerol's high boiling point (290°C). It was believed that changing to a binary solvent mixture would allow for tuning of the rheological behaviour of the inks, as well as lowering the boiling point.

Figure 7a and b show that changing to a binary mixture (ink #2, glycerol – water 80w/20w) introduces some thixotropy. Using either pure propylene glycol or a 90w/10w mixture of propylene glycol and water also results in thixotropy. Thixotropy seems more pronounced in the water containing vehicles, although differences are relatively small and thixotropic effects become insignificant already at relatively low shear rates (i.e. > 2.4 s^{-1}). The propylene glycol-(water) and glycerol-water inks show a similar degree of pseudoplasticity to ink #1, but the SEM micrograph in Figure 9 shows that possibly due to their thixotropic behaviour, albeit minor, they produce layers of more uniform thickness. Enhancing the thixotropy in inks #2-4 may result in further improvements of the prints. It can currently not be discounted that the observed thixotropy in inks #2-4 is at least partly due to an increase in the binder content as compared to ink #1. Also, the propylene glycol (-water) based inks generally result in better prints than the glycerol based inks, which could not be predicted on the basis of the rheological data presented here alone. More studies are necessary to elucidate the interaction of the binder (PVP) with the various solvent molecules, i.e. water, glycerol and propylene glycol, to fully understand the differences in printing characteristics. Polar systems are known for unusual dipole interactions as well as hydrogen bonding. Interesting rheological as well as interfacial (i.e. wetting, adhesion, etc.) properties have been reported in similar polymer – solvent systems, due to these interactions[8, 9]. This will be the focus for further study. The different molecular interactions may also be responsible for the increased viscosity of ink #4 (PG-water 90/10) as compared to ink #3 (PG/water 100/0). This seems unusual considering the results presented in Figure 3 and Figure 5. However, the addition of solids to the vehicles introduces a new interaction. Water is known to affect the surface chemistry of oxides strongly due to charge effects. More indepth experiments are needed here too to study this effect. The addition of water may also affect the quality of the dispersion negatively, causing some minor agglomeration of the YSZ particles. Agglomeration is known to cause an increase in viscosity due to the trapping of solvent in the interstitial spaces within the agglomerates.

Co-firing anode supports and printed electrolytes

A non-optimised two step firing technique was utilised to co-fire green anode tapes with printed electrolytes. This resulted in flat ceramics, but with varying degrees of porosity in both electrolyte and anode support. More studies are required to obtain a consistent firing protocol, which results in both porous anode supports as well as dense electrolyte layers. Progress will hopefully be reported in future papers.

CONCLUSIONS

New screen printing inks with increased polarity as compared to conventional organic inks have been developed and proven to be compatible with green aqueous tapes. The presence of thixotropy in some of the used inks is believed to be beneficial in obtaining uniformly thick prints. Best printing results were obtained for inks based on propylene glycol with PVP binder.

ACKNOWLEDGEMENTS

The research leading to these results has received funding from the Fuel Cells and Hydrogen Joint Undertaking under grant agreement n° 256730.

REFERENCES
1.   M. C. Verbraeken; B. Iwanschitz; A. Mai; J. T. S. Irvine, *J. Electrochem. Soc.,* **159** (11), F757 (2012).
2.   M. Cassidy; S. Boulfrad; J. T. S. Irvine; C. Chung; J. M.; M. C.; S. Pyke, *Fuel Cells,* **9** (6), 891 (2009).
3.   P. Gordes; N. Christiansen; E. J. Jensen; J. Villadsen, *J. Mater. Sci.,* **30** (4), 1053 (1995).
4.   G. J. Wright; J. A. Yeomans, *Int J Appl Ceram Tec,* **5** (6), 589 (2008).
5.   I. W. Chen; X. H. Wang, *Nature,* **404** (6774), 168 (2000).
6.   P. N. Shankar; M. Kumar, *P Roy Soc Lond a Mat,* **444** (1922), 573 (1994).
7.   T. F. Sun; A. S. Teja, *J. Chem. Eng. Data,* **49** (5), 1311 (2004).
8.   M. T. Islam; N. Rodriguez-Hornedo; S. Ciotti; C. Ackermann, *Pharm. Res.,* **21** (7), 1192 (2004).
9.   L. Li; V. S. Mangipudi; M. Tirrell; P. A.V., Direct measurement of surface and interfacial energies of glassy polymers and PDMS. In *Fundamentals of Tribology and Bridging the Gap Between the Macro- And Micro-Nanoscales,* Bhushan, B., Ed. Springer: 2001; pp 305.

MODIFICATION OF SINTERING BEHAVIOR OF Ni BASED ANODE MATERIAL BY DOPING FOR METAL SUPPORTED-SOFC

Pradnyesh Satardekar[1], Dario Montinaro[2], Vincenzo M. Sglavo[1]
1. Department of Industrial Engineering, University of Trento, via Mesiano 77, I-38123 Trento, Italy
2. Sofcpower Spa, Viale Trento, 115/117 – c/o BIC – modulo D, I-38017 Mezzolombardo, Italy

ABSTRACT
The modification of the reduction kinetic of NiO and the interaction between the anode and steel during the fabrication of Metal Supported Solid Oxide Fuel Cells (MS-SOFC) is studied in the present work. With the aim to limit NiO reduction under inert atmosphere at high temperature, doping elements such as Al and Ce were considered for NiO powders modification and anode production. In order to simulate the reactions at the metal/anode interface, NiO/YSZ/steel composites were prepared with pure and Al-doped NiO. A sudden volume expansion above 1000^0C followed by substantial shrinkage above 1200^0C was observed for the composites when sintered in Ar at 1400^0C. Such volume expansion can be related to the oxidation of steel due to the RedOx reaction between NiO and steel. Moreover, it was found that the volume expansion, i.e. the steel oxidation, can be minimized to a good extent when Al-doped NiO is used. Hence it is proposed that Al-doped NiO is a promising candidate material to be used for anodes in high temperature sintering of MS-SOFC.

INTRODUCTION
The reduction in the operating temperature from 850^0C-1000^0C to 700^0C for conventional ceramic Solid Oxide Fuel Cell (SOFC) has opened the possibility to use less expensive stainless steels not only for interconnects and BoP, but also for the SOFC support.

Metal Supported-Solid Oxide Fuel Cells (MS-SOFC) represent a promising new design for fuel cells which may overcome the limitations of anode-supported cells (such as poor thermal cycling resistance and brittleness, Nickel phase re-oxidation upon exposure to transient uncontrolled conditions) due to the much better mechanical properties of the support that is represented by a porous thick metal substrate , the thickness of the ceramic layers (anode/electrolyte/cathode) being in the order of 10-50 μm, only. In addition, in this design (Fig. 1), the replacement of the thick Ni/YSZ cermet with ferritic stainless steel leads to several benefits in term of fabrication cost and safety.

MS-SOFCs are not only cost efficient, but they also promise other advantages such as robustness and tolerance to rapid thermal- and redox-cycling. Ferritic stainless steel is a widespread candidate for the support, because it is cheap, its thermal expansion coefficient matches closely with that of metal interconnects and it has a good high temperature oxidation resistance[1].

SOFC fabrication normally requires high sintering temperatures (i.e. 1300-1400°C) in order to achieve a perfect gas-tightness of the YSZ electrolyte. In the case of MS-SOFC, in order to prevent steel oxidation, sintering steps have to be performed under inert or protective atmosp-

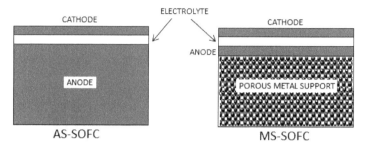

Fig.1. Comparison of design of Anode Supported SOFC and Metal Supported SOFC

here. The combination of high temperature and controlled atmosphere lead to several issues when co-sintering of metal-anode-electrolyte multilayers is considered as a production route. Such conditions, in fact, results in NiO reduction and extensive coarsening of Ni phase, in addition to significant interdiffusion of Ni, Cr and Fe between the anode and the porous metal support [1, 2].

The interdiffusion of species at the metal/anode interface can be limited by the addition of a diffusion barrier layer between the metal and the anode. Such extra-layer is often made by CeO_2 due to its low reactivity with the adjacent layers and its good electrical conductivity (0.89 S/cm at 800^0C) under typical anode operating condition [2]. Nevertheless, although inhibiting solid state interdiffusion, CeO_2 alone is not sufficient to prevent NiO reduction by the chromium contained in the steel, because the oxygen partial pressure in the furnace can be controlled by the release of O_2 due to NiO reduction and, therefore, it does not need a direct interface between the two phases. For this reason, while interdiffusion of Fe, Cr and Ni species can be prevented by the addition of the barrier layer, Cr-NiO RedOx reaction and Ni coarsening still represents a problem to solve. In order to overcome these processing problems, several approaches, in which the electrolyte and/or the anode are fabricated in separate steps and at different temperatures, have been proposed. A reduction of the electrolyte sintering temperature was demonstrated to be efficiently achieved by low-temperature processes such as pulsed laser deposition (PLD), plasma spray and electrophoretic deposition (EPD) [2, 3, 4]. In this case, the anode layer is never exposed to very high temperatures and, therefore, the interdifussion is significantly prevented. Other multi-step approaches used to meet the challenge for high temperature anode processing are represented by the fabrication of metal-electrolyte substrates that, after sintering the electrolyte at high temperature, are impregnated with the Ni catalyst, thus overcoming the interdiffusion reactions occurring at high temperature [5, 6]. However, these low temperature routes often involve expensive equipment and, due to the multi-step approach, are less attractive for the large scale productions.

Cost efficient single step co-sintering of metal/anode/electrolyte multilayer laminate is a very attractive route for MS-SOFC fabrication.

Reduction and oxidation kinetic of metal oxides involve the diffusion of metal cations and oxygen anions through the oxide lattice and along high diffusivity pathways, such as grain boundaries. The contribution of different transport mechanisms depends on temperature [7]. The RedOx behavior of nickel oxide was found to be tunable by the addition of reactive elements (RE), usually rare-earths, which have a high affinity for oxygen [7, 8, 9]. Reactive element addition to Ni limits NiO scale growth probably operating by inhibiting grain boundary transport. On the other hand, RE additions may increase the oxidation rate of Ni at higher temperatures (1100-1200°C) [8]. Additions of RE to improve the RedOx tolerance in SOFC anodes have been proposed by Larsen et al. [10, 11]. In addition, Cologna et al. demonstrated that the sintering profile of the anode material can be controlled by the addition of Al_2O_3 and CeO_2 and high quality flat cells can be obtained by reducing the sintering mismatch among the different ceramic layer [12]. Tikekar et al. studied the reduction and re-oxidation kinetics of Ni/YSZ anode material and demonstrated that small amount of stable oxides like CaO, MgO and TiO prevent Ni re-oxidation under transient operating conditions [13]. Despite several findings on the beneficial effect of doped-NiO on the anode RedOx behavior under SOFC operating conditions (600-800°C), no investigations have been so far performed on the effect of doping elements on the reduction and coarsening mechanism of NiO upon the high temperature and inert atmosphere sintering conditions.

In the present work, Al- and Ce-doped NiO powders were considered for anode preparation in MS-SOFC. The effect of doping on the reduction of NiO and anode/metal interdiffusion were investigated. Preliminary investigations were carried out considering dense Crofer steel as substrate while a commercial ferritic stainless steel was considered for composites characterization and half-cell production. Mutilayers of YSZ/NiO-YSZ/CeO$_2$/steel were prepared by tape casting with pure and doped NiO, which were then sintered under controlled conditions. Ni/8YSZ cermet and 8YSZ were used for the anode and the electrolyte, respectively and, CeO$_2$ was used for diffusion barrier layer. Composites of NiO-YSZ-steel powders were prepared in order in investigate the anode steel interaction.

MATERIALS AND METHODS

Doped powders preparation
Al(NO$_3$)$_3$.9H$_2$O (Fluka, Germany) was used as precursor to produce 0 to 5 wt% Al-doped NiO. NiO (J.T.Baker, USA), Al(NO$_3$)$_3$.9H$_2$O and water were mixed for 20 h in a plastic jar containing zirconia balls (Inframat Advanced Materials, USA) and then dried at 110^{0}C overnight. The powder was manually ground in mortar, calcined at 900^{0}C for 10 h and then again manually ground to avoid any agglomerate. Ferritic stainless steel powder containing 22 wt% Cr, supplied by Höganäs AB, was also considered. These powders were then used to produce the NiO-8YSZ-steel (1:1:2 wt%) composites by energetic mixing (Turbula, Switzerland) in a plastic jar containing zirconia balls and water for 6 h. The powders were again dried overnight at 110^{0}C and then manually ground.

Ce-doped NiO powder was produced in similar way for which Ce(NO$_3$)$_3$.6H$_2$O (Alfa Aesar, Germany) was used as precursor.

Preliminary investigations

For the initial studies, anode ink composed of NiO/YSZ with pure, Al- and Ce- doped NiO was prepared and screen printed on the dense Crofer 22 APU substrates (1 mm thick). Samples were then heated at 2^0C/min up to 1400^0C and sintered for 2 h under Ar atmosphere.

In order to simulate the volume expansion occurring at the anode/metal interface, composite materials were analyzed by dilatometry (Linseis GmbH, Germany) under the sintering conditions typical for MS-SOFC fabrication. Specimens were prepared by pressing 2.5 g of the composite powders in a 20 mm diameter circular dye at 100 MPa for 2 min, reaching the thickness of about 2 mm. The resulting pellets were then cut into rectangular bars with size 12 x 5 x 2 mm^3 and used for the dilatometry. The used heating profile was 2^0C/min up to 1400^0C in Ar condition at flow rate of 400 ml/min.

In order to evaluate the effect of each material on the composite mixture, dilatometric analyses were also performed on single-phase specimens containing only steel and YSZ subjected to same milling conditions used for NiO-YSZ-steel composite.

X-ray diffraction measurements were performed by a Rigaku Geigerflex X-ray Powder Diffractometer (Japan) on the composites heat treated in the dilatometer under the same heating rate and Ar flow rate to investigate the phase transformations taking place in NiO-8YSZ-steel composite. Some samples were also prepared by heating in the dilatometer and stopping the cycle at temperatures corresponding to selected discontinuities observed in the dilatometric curve and XRD measurements were taken on crushed sample after cooling down at room temperature. XRD scanning was carried out between 25^0 to 85^0 with step interval of 0.05^0 and hold of 8 s for each step.

Half-Cells production

Multilayered half cells (YSZ/NiO-YSZ/CeO$_2$/steel) were prepared by water-based tape casting technology. For the anode, a ratio of 1:1 was used for NiO and YSZ and the half-cell was produced with pure and doped NiO. Vynil- and acrylic-base emulsions were used as binders to prepare the slurries for the steel and ceramic layer, respectively.

The microstructure of the composite pellets and multilayer half-cells was observed by scanning electron microscope (JEOL JSM 5500, Japan).

RESULTS AND DISCUSSION

Anode-steel interaction

The dilatometric plot in Fig. 2 shows the behavior of the samples heat treated under Ar and air atmosphere. When pure steel sample is heat treated in air, it shows a volume expansion obviously related to oxidation, but when heated in Ar it shows a conventional sintering behavior with a shrinkage onset at 1170^0C. This result allows to exclude that the volume expansion of the steel could be related to oxidation due to oxygen impurities in the Ar atmosphere. On the other hand, it can be observed that NiO/YSZ/Steel composite shows a sudden volume expansion above 800^0C followed by a substantial shrinkage after 1200^0C in Ar. This volume expansion in the composite heated in Ar can be attributed to the oxidation of steel due to the redox reaction between NiO and steel. It is well known from the Ellingham diagram [14] that Fe and Cr oxides are thermodynamically more stable than NiO; therefore, there is a thermodynamically favored oxidation of steel due to the redox reaction between steel and NiO, producing Ni. From these results it is proposed that steel-NiO RedOx reactions involving volume changes at the interface between Ni-based anode and the metal substrate could be the key sources for major issues like coarsening, delamination and cracking of MS-SOFC. As shown in Fig.2, YSZ shows the onset of sintering at around 1200^{0}C accounting for shrinkage in the composite above 1200^{0}C. It is worth to notice that the net shrinkage of the NiO/YSZ/steel sample, at 1400°C, is almost zero, such value being definitely not suitable for cells fabrication, where at least 20% volume reduction is required for the co-sintering with the electrolyte. It is also interesting to observe the two slight rate changes in the volume increase for the pure steel when fired under Air atmosphere. This result suggests a change of the oxidation mechanism of the steel as a function of the temperature and it could be related to the formation of Cr-Fe oxides with different structure, on the metal particle surface. Therefore, the change of slope in the curve suggests 3 ranges of steel oxidation, between 800°-1050°C, 1050°-1150°C and above 1150°C.

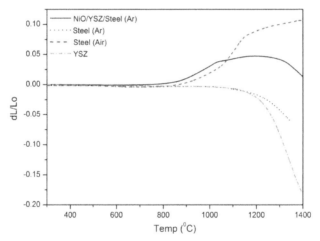

Fig.2. Change in relative length (dL/L$_0$) as a function of temperature for pure materials and composite under different atmospheres.

It is observed in Fig.3 (and also Fig.2) that the sintering curve for NiO/YSZ/steel composite follows a non-monotonic behavior, displaying a step at about 1000°C, a maximum volume expansion at ≈1160°C and a continuous behavior up to 1400°C. As described before, XRD measurements of as prepared NiO/YSZ/steel composite and composites from dilatometer were taken after heat treating at such characteristic temperatures and then cooling down to room temperature. The XRD pattern related to these three samples is shown in Fig.4. For the as-prepared sample, the XRD pattern of the composite shows NiO peaks; the intensity of these peaks decreases when the sample is fired up to 1000^{0}C and then disappears at 1160^{0}C. It should be noted that peaks corresponding to Ni appear at 1000^{0}C, this pointing out that, at such temperature, NiO already starts to reduce, producing Ni. The peaks corresponding to Fe$_2$O$_3$ are also observed at $1000^{0}C$ and, as expected, the peaks intensities for steel decreases as Fe_2O_3 increases with temperature. At $1160^{0}C$, the peaks for Ni decreases and there is the appearance of new Fe-Cr-Ni phase. We can speculate that Ni, produced from the redox reaction, rapidly diffuses into the steel to form the new Fe-Cr-Ni phase or mixed oxides, leading to the disappearance of the Ni phase. Conversely, no single Cr_2O_3 phase was observed. However, due to the low intensity of the peaks, an unambiguous identification of the minor phases was not achieved.

These findings are also supported by EDXS maps (Fig.5) recorded on the three samples. Almost no interdiffusion was observed at 1000°C, where only the formation of a thin Cr-rich scale is clearly visible on the surface of metal particles. This scale increases in thickness when the sample is heated up to 1160°C but, at this temperature, some spherical Ni-rich particles are observed, together with a slight Fe diffusion in the ceramic phase. At 1400°C, the extent of Fe and Cr outward diffusion from the steel to the ceramic matrix is huge, forming a thick layer containing Cr, around the metal particle. From the SEM images, since the edges of the metal particle are quite sharp, this Cr-rich phase seems to be not an oxide scale adherent to the metal particle but a mixed new phase formed through the ceramic network. It is also interesting to observe that, while Cr diffusion is limited to this reaction layer, Fe is homogeneously distributed in the ceramic matrix, forming alloys and/or mixed oxides with Ni. On the other hand, Ni diffusion through the steel is not significant enough to be recorded by EDXS maps. We could speculate that the fast formation of the Cr scale on the steel surface, already below 1000°C, prevents further Ni diffusion in the steel, even at high temperature. However, such Cr-rich scale, is not able to block Fe outward diffusion or, being composed of a mixed Cr-Fe oxide, it releases Fe in the ceramic phase as the temperature increases above 1000°C.

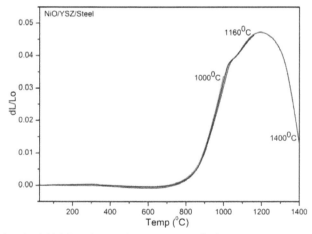

Fig.3. Change in relative length (dL/Lo) as a function of temperature for NiO/YSZ/steel composite when heat treated in Ar. The three curves correspond to three different samples heated up to each selected temperature and then cooled down and used for XRD analysis.

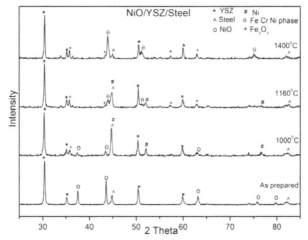

Fig.4. XRD pattern of NiO/YSZ/steel composite samples after heating at different temperatures in Ar.

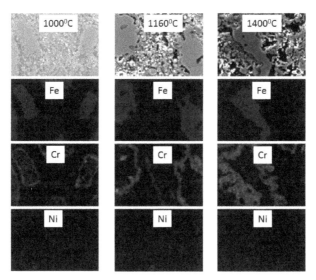

Fig.5. EDXS maps of NiO/YSZ/steel composite samples after heating at different temperature in Ar.

Effect of doped-NiO

In order to limit the steel-induced NiO reduction and, consequently, the associated sudden volume change, NiO redox kinetic was modified by the addition of reactive doping element, Al and Ce, to NiO powder. A preliminary investigation was carried out by printing

NiO/8YSZ layers on dense Crofer 22APU plates and sintering under Ar atmosphere, 2 h at 1400°C. The microstructure of samples with pure-NiO, 2wt%Al-NiO and 3wt%Ce-NiO after sintering is shown in Fig.6. This preliminary test shows that, in the case of pure NiO, the adhesion between substrate and anode material is not uniform, with the formation of cavities at the interface, probably due to Cr evaporation and diffusion through the ceramic layer. In addition, several Cr-rich oxide grains, probably related to a mixed Ni-Cr-Fe spinel phase, were observed in the bulk of the anode layer. On the other hand, in the case of Al-doped-NiO, the anode shows good adhesion to the substrate, with a limited occurrence of voids at the interface. Similar results were observed in the case of Ce-doping although the adhesion between anode and substrate was not as good as in Al doped samples. An increased porosity was observed in the case of Al-NiO, while both pure- and Ce-NiO containing layers showed similar microstructural density.

Composite mixtures of steel/8YSZ with doped and undoped NiO were prepared and sintered at 1350°C under Ar atmosphere. The resulting microstructure of the samples is shown in Fig.7. The sample prepared with pure NiO showed the coarsest grain structure, due to the extent of densification at high temperature. On the other hand, the two samples prepared with Al and Ce doping retain a quite fine ceramic microstructure, the composite with Al-NiO possessing the finest structure (Fig.7b). In addition, large metal grains are distinctly visible only in the case of Al-NiO, while they appear in a continuous network with the ceramic phase in pure and Ce-doped samples. On this basis, the successive investigation was focused on Al-NiO only.

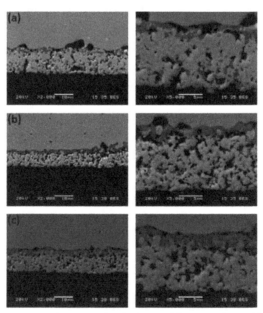

Fig.6. SEM images showing the microstructure between anode and substrate at different magnification. (a) Pure NiO; (b) 2 wt% Al doped NiO; (C) 3 wt% Ce doped NiO.

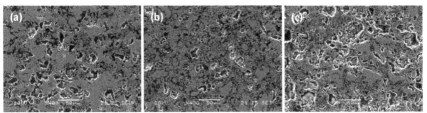

Fig.7.Microstructure of NiO/8YSZ/steel composite after sintering at 1350°C for 2 h under Ar atmosphere. Composite were prepared by using: (a) pure NiO; (b) Al-doped NiO; (c) Ce-doped NiO.

In order to study the effect of increasing dopant concentration on the consolidation behavior and to optimize the NiO powder composition, composites containing increasing amounts of aluminum were prepared and analyzed at the dilatometer. The sintering temperature for such experiment was set to 1350°C (under Ar atmosphere) to propose a cell production process involving less severe sintering conditions. Results are given in Fig. 8. The shrinkage of the composites drastically changes up to 3 wt% Al doping, while the increase of Al content from 3 to 5 wt% does not significantly change the sintering profile. The occurrence of the step at about 1000°C is still unclear, but probably it does not depends to the dopant since this behavior was observed even for undoped NiO. It is worth to note that Al-doping up to 3 wt%, significantly reduces the volume expansion in the composites, leading to a larger final shrinkage.

On this basis, doped NiO powder (3 wt%Al-NiO) was selected for MS-SOFC fabrication. The cell was produced by tape casting aqueous powder suspensions, the architecture of the cell being 8YSZ/NiO-3Al+8YSZ/CeO₂/steel. A cross-section of this cell is shown in Fig.9. Probably due to the short sintering time, the densification of the electrolyte was not fully achieved for this test. However, despite the high temperature, a very porous anode microstructure was observed.

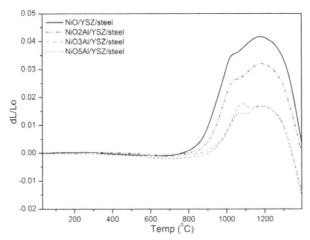

Fig.8. Change in relative length as a function of temperature for NiO/YSZ/steel composite with pure and Al-doped NiO.

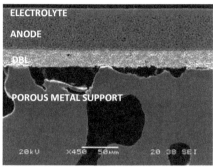

Fig.9. Cross-section of co-sintered 8YSZ/Al-NiO-8YSZ/CeO2/steel half-cell sintered for 2 h at 1350°C under Ar atmosphere.

CONCLUSIONS

High temperature co-sintering of metal/ceramic layers under inert atmosphere, in MS-SOFC, leads to problems associated to volume change and different shrinking behaviour of the layers, resulting in Ni-coarsening, cracking and delamination of the electrodes.

In the present work, steel oxidation by oxygen provided by nickel oxide was identified as the main source for volume expansion and the limited shrinkage of the half-cell multilayer. In addition, intense Cr-Fe-Ni interdiffusion and Ni coarsening were observed when the different materials are co-sintered. The Cr oxide formed on the steel surface at 1000°C seems to limit Ni diffusion through the steel, but it cannot prevent Fe outward diffusion when the temperature is further increased.

It was found that the addition of doping elements (Al and Ce) to NiO significantly reduces steel oxidation and volume expansion, probably affecting the NiO reduction rate. This work was focused on Al-doping and it was observed that aluminum additions up to 3 wt% reduce the volume expansion.

ACKNOWLEDGEMENTS

D. Montinaro is grateful to the European Commission for financial support through the FP7 FCH-JU 2009-1 RAMSES project (Nr 256768).

REFERENCES

[1] M. C. Tucker, "Progress in metal-supported solid oxide fuel cells: A review," *Journal of Power Sources*, vol. 195, pp. 4570-4582, 2010.

[2] M. Brandner, J. Froitzheim, M. Bram, H. P. Buchkremer and D. Stover, "Electrically conductive diffusion barrier layers for Metal-Supported SOFC," *Solid State Ionics*, vol. 179, pp. 1501-1504, 2008.

[3] S. Hui, D. Yang, Z. Wang, S. Yick, C. Deces-Petit, W. Qu, A. Tuck, R. Maric and D. Ghosh, "Metal-supported solid oxide fuel cell operated at 400-600C," *Journal of Power Sources,* vol. 167, pp. 336-339, 2007.

[4] N. P. Brandon, A. Blake, D. Corcoran, D. Cumming, A. Duckett, K. El-Koury, D. Haigh, C. Kidd, R. Leah, G. Lewis, C. Mathews, N. Maynard, N. Oishi, T. McColm, R. Trezona, A. Selcuk, M. Schmidt and L. Verdugo, "Developement of metal-supported solid oxide fuel Cells for operation at 500-600 C," *Journal of Fuel Cell Science and Technology,* vol. 1, pp. 61-65, 2004.

[5] M. C. Tucker, G. Y. Lau, C. P. Jacobson, L. C. DeJonghe and S. J. Visco, "Stability and robustness of metal-supported SOFCs," *Journal of Power Sources,* vol. 175, pp. 447-451, 2008.

[6] M. C. Tucker, G. Y. Lau, C. P. Jacobson, L. C. DeJonghe and S. J. Visco, "Performance of metal-supported SOFCs with infiltrated elecrode," *Journal of Power Sources,* vol. 171, pp. 477-482, 2007.

[7] R. Haugsrud, "On the high temperature oxidation of nickel," *Corrosion Science,* vol. 45, pp. 211-235, 2003.

[8] D. P. Moon, "The Reactive Element Effect on the Growth Rate," *Oxidation of metals,* pp. 47-66, 1989.

[9] P. H. Larsen, C. Chung and M. Mogensen.United States Patent US 2008/0166618 A1, 2008.

[10] Larsen and e. al.United States Patent US 2008/0188818 A, 2008.

[11] A. R. Contino, M. Cologna, S. Modena and V. M. Sglavo, "Effect of Doping Elements on the Redox Kinetics of NiO-YSZ Powders for SOFC Applications,," in *ECS Transections,* 2009.

[12] M. Cologna, A. R. Contino, D. Montinaro and V. M. Sglavo, "Effect of Al and Ce doping on the deformation upon sintering in sequential tape cast layers for solid oxide fuel cells," *Journal of Power Sources,* vol. 93, p. 80–85, 2009.

[13] N. M. Tikekar, T. J. Armstrong and A. V. Virkar, "Reduction and Reoxidation Kinetics of Nickel-Based SOFC Anodes," *Journal of The Electrochemical Society,* vol. 153, pp. A654-A663, 2006.

[14] H. Ellingham, *J Soc Chem Ind (London),* vol. 63, p. 125, 1944.

NICKEL PATTERN ANODES FOR STUDYING SOFC ELECTROCHEMISTRY

HC Patel *, V Venkataraman, PV Aravind
Delft University of Technology, Process & Energy Laboratory, Section Energy Technology,
Leeghwaterstraat 44, 2628 CA Delft, The Netherlands
*Corresponding Author
Email address : H.C.Patel @tudelft.nl

ABSTRACT

The SOFC anode is a complex 3D structure where the gas phase, ionic phase and electronic phases are intermingled. This not only makes it difficult to quantify the triple phase boundary (TPB) length exactly, but also makes individual processes like charge transfer, diffusion, surface diffusion etc, difficult to localise and study. Pattern electrodes offer the advantage of having a well defined TPB length, so that reactions are localised. Impedance spectroscopy on pattern electrodes can provide mechanistic information in addition to identifying rate limiting steps. In this work nickel is sputtered on YSZ with DC magnetron sputtering using a Nickel target and a metallic mask on YSZ substrate in a specific pattern to give a well-defined TPB length. A TPB length of 0.2027 m/cm2 was achieved. This method is much easier and is less expensive than the photolithographic techniques used earlier (Bieberle et al ETH Zurich, Mizuzaki et al Yokohama national University and Boer de et al University of Twente). Impedance spectroscopy is carried out on symmetrical cells in pure hydrogen with 4.3 % water vapour. Preliminary results from equivalent circuit fitting of the impedance spectra indicate that the results are comparable to results available in literature and the rate limiting mechanisms can be identified.

INTRODUCTION

SOFCs have the potential of being highly efficient producers of energy in a sustainable manner. Cermet of Nickel (Ni) - Yttria stabilized Zirconia (YSZ) or Gadolina doped Ceria (GDC) are now commercially used as SOFC anodes. The cermet anode is a 3D structure with interpenetrating ionically conducting phase (YSZ) and an electronic conducting phase (Ni). As a result of this mixing of the two phases the microstructure is difficult to quantify exactly. Moreover, while operating the microstructure may undergo changes. Thus the reaction mechanisms are difficult to elucidate without accurate knowledge of the triple phase boundaries (TPBs) – where the ionically conducting phase, electronically conducting phase and gas phase are in contact with each other. The knowledge of reaction mechanisms at the TPBs is essential for understanding the fuel oxidation and refining existing designs. One way to localize and study reactions is to make anodes with well defined geometry and hence TPBs. Also variations across the third dimension (thickness) can be omitted to provide a much simpler geometry with well defined TPB lengths. Hence a 2D pattern is chosen where the TPB length can be measured accurately. Impedance measurements on this geometry can give invaluable insight into the reaction mechanisms at the SOFC anode. Previous studies in making pattern anodes have utilised lithography and etching to make patterns with well defined TPB lengths [1,2,3]. In this work, we use a metallic mask instead, which is inherently much more accessible and cheaper than using semiconductor based technologies.

EXPERIMENTAL

A stainless steel mask was fabricated in-house with a desired pattern that could house the electrolyte discs. Ni was then sputtered by DC magnetron sputtering. To ensure the structural stability of the Ni pattern at SOFC operating temperatures it was necessary to deposit a thickness of at least 800 nm [4]. The actual thickness of the Ni pattern deposited for this experiment was 1780 nm. The sputter parameters used in this process is listed in *Table 1*. It must be pointed out here that a symmetrical anode cell was sputtered with identical patterns on both sides of the YSZ disk.

Table 1: Parameters for sputtering

DC Bias	No bias applied
Sputter Gas	Argon 99.999%
Basis pressure in the chamber	$5*10^{-7}$ mbar
Pressure during sputtering	$3*10^{-6}$ bar
Size of the target	2 inch
Power at the target	100-150 W
Rate of sputtering	0.12nm/sec

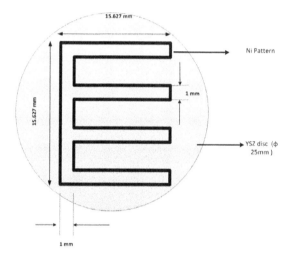

Figure 1: Nickel pattern design with dimensions (not to scale)

a)

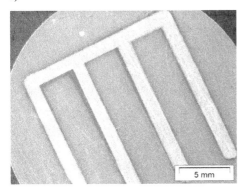

b)

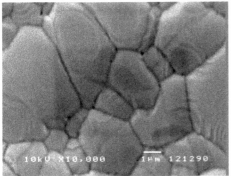

Figure 2 Nickel pattern as deposited a) Light microscopy image of the pattern. b) SEM image of the pattern

Impedance measurements were carried out on the pattern electrodes starting at 850° C and subsequently at lower temperatures with 50°C temperature intervals. Only the results obtained for $700\ ^0$C are discussed here. The measurements were carried out using the setup described by Aravind et al [6]. The cell was allowed to stabilize at each temperature for 1 hour before measurements were recorded. Impedance measurements were carried out in the frequency range 10^5 Hz to 0.01 Hz. The fuel was 96% hydrogen humidified to around 4% moisture.

RESULTS AND DISCUSSION

The sputtering through steel masks gave excellent 2D patterns but at the cost of higher line widths of Ni. The line width obtained was 1mm as compared to micrometer range obtained by other researchers [1, 2,3]. The TPB length obtained was 150 mm. From light microscopy images it can be seen that there are no defects and the pattern is continuous. Also Scanning Electron Microscopy (SEM) images showed that the pattern is not porous and the TPB estimated will be along the boundary of the pattern

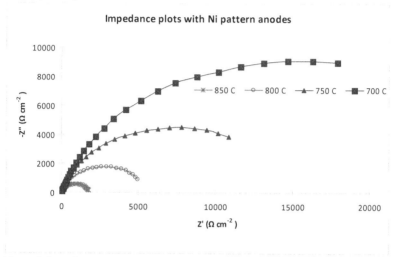

Figure 3: Impedance spectra with Nickel Pattern anodes

Figure 3 shows the impedance spectra obtained with Nickel pattern anodes at various temperatures. For this work only the total electrode conductivity is used i.e. the intercept with the real axis and hence equivalent circuit fitting is not presented. The resistance obtained is very large as compared to what is expected from SOFC anodes [5] [6] but here the TPB length is smaller as well.

The values of the electrode conductivity from the experiments conducted with Ni pattern electrodes were compared with the electrode conductivity values obtained by other research groups who have conducted similar experiments with Ni pattern electrodes as well. However the main difference is that a model Ni pattern electrode with only one TPB length (0.2027 m/cm^2) was used in this work as compared to model Ni pattern electrodes with a range of TPB lengths in experiments performed by other research groups. Hence there was a need to interpret and scale the results, to the TPB length of 0.2027 m/cm^2, from other research groups for comparison purposes. Also, the partial pressure of water was different and the conductivity was scaled according to the dependence of impedance on water partial pressure as presented by various authors. Reasonable assumptions were made to calculate these values.

Figure 4 shows the electrode conductivity values obtained in this work in comparison to Bieberle et al, Mizusaki et al and De Boer at 2.33 kPa partial pressure of water, rest hydrogen, L_{TPB} of 0.2707 mcm^{-2} and 700 ^{0}C. As seen from *Figure 4*, there is quite a spread in the data obtained by various groups but we can conclude that the total electrode conductivity value obtained in this work for similar experimental conditions is within a reasonable range. This shows that model 2D electrodes prepared in a much simpler way yield relatively good results, which are quite comparable with the expensive fabrication techniques. The conductivity obtained is lower than that by Bieberle et al and Mizusaki et al but is higher than De Boer even though similar methods for making the pattern was used by all three.

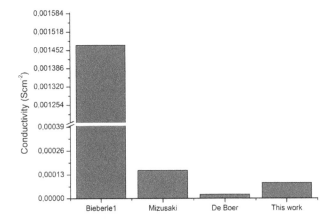

Figure 4: Comparison of electrode conductivity with different groups at similar conditions

A symmetrical cell Nickel pattern anode was made using a metallic mask. It is expected that the TPB line is not very sharp because of the imperfect contact between the mask and electrolyte. This would probably lead to some error in estimation of the TPB length. However, when compared to results obtained in previous work it is found that the conductivity is within reasonable limits. Fitting of this data to equivalent circuits and elucidating limiting processes is considered as part of future work on this.

CONCLUSIONS

A Nickel pattern anode was fabricated using much cheaper methods than conventionally used. The TPB length achieved was 0.2707 m/cm2. Comparing with previous work in this area, it is found that the conductivities obtained are similar. It is recognized here that the TPB is likely to be more diffuse in this work, but since the conductivity is large enough it is acceptable. Further analysis on the impedance spectra obtained with detailed modeling should help with understanding the reaction mechanisms of fuel oxidation at the SOFC anode.

ACKNOWLEDGEMENTS

The authors would like to thank Ing. Herman Schreuders from the Department of Chemical Engineering, Faculty of Applied Sciences Technical University of Delft for help with the deposition.

REFERENCES
[1] Bieberle, A., L.P. Meier, and L.J. Gauckler, *The Electrochemistry of Ni Pattern Anodes Used as Solid Oxide Fuel Cell Model Electrodes.* Journal of The Electrochemical Society, 2001. **148**(6): p. A646-A656.
[2] Mizusaki, J., et al., *Kinetic studies of the reaction at the nickel pattern electrode on YSZ in H_2-H_2O atmospheres.* Solid State Ionics, 1994. **70/71**: p. 52-58.

[3] Boer, B.d. PhD Thesis, *Hydrogen oxidation at porous nickel and nickel/yttriastabilised zirconia cermet electrodes*, in *Inorganic Materials- Faculty of Chemical Engineering*1998, University of Twente ISBN 90-36511909.
[4] Bessler WG, Vogler M, Stormer H, Gerthsen D, Utz A, Weber A and Ivers-Tiffe E *Model anodes and anode models for understanding the mechanism of hydrogen oxidation in solid oxide fuel cells* Phys. Chem. Chem. Phys., 2010, **12**, 13888–13903
[5] Aravind, P.V, Phd Thesis, Technical University of Delft 2007 ISBN-13: 978-90-9022534-0.
[6] Aravind, P.V., J.P. Ouweltjes, and J. Schoonman, *Diffusion Impedance on Nickel/Gadolinia-Doped Ceria Anodes for Solid Oxide Fuel Cells.* Journal of The Electrochemical Society, 2009. **156**(12): p. B1417-B1422.

ASSESSMENT OF $Ba_{1-x}Co_{0.9-y}Fe_yNb_{0.1}O_{3-\delta}$ FOR HIGH TEMPERATURE ELECTROCHEMICAL DEVICES

Zhibin Yang[1, 2], Tenglong Zhu[1], Shidong Song[1], Minfang Han[1]*, Fanglin Chen[2]**

1. Union Research Center of Fuel Cell, China University of Mining & Technology, Beijing, 100083, P.R. China
2. Department of Mechanical Engineering, University of South Carolina, Columbia, SC 29208, USA
*hanminfang@sina.com; **chenfa@cec.sc.edu

ABSTRACT

Mixed-conducting perovskite-type materials can be used in many applications such as Oxygen Transport Membranes (OTMs), Solid Oxide Fuel Cells (SOFCs) and Solid Oxide Electrolysis Cells (SOECs). Recently, it has been reported that introduction of Nb in the B-site of cobalt-based perovskites can significantly improve chemical stability and enhance performance. In this study, $Ba_{1-x}Co_{0.9-y}Fe_yNb_{0.1}O_{3-\delta}$ has been introduced based on the study of calculate analysis- experimental data. Further, the performance and stability of $Ba_{1-x}Co_{0.9-y}Fe_yNb_{0.1}O_{3-\delta}$ (x=0-0.15, y=0-0.9) for high temperature electrochemical device applications such as OTM, SOFC and SOEC has been introduced. The results show that $Ba_{0.9}Co_{0.7}Fe_{0.2}Nb_{0.1}O_{3-\delta}$ can be a very promising candidate for the cathode of SOFCs and membrane materials of OTM application. Maximum power density of 1.1 $W \cdot cm^{-2}$ was obtained for $La_{0.8}Sr_{0.2}Ga_{0.83}Mg_{0.17}O_{3-\delta}$ (LSGM) electrolyte supported cells with $Ba_{0.9}Co_{0.7}Fe_{0.2}Nb_{0.1}O_{3-\delta}$ as cathode and Ni-GDC as anode operated at 800 °C. Under an ambient air / helium oxygen gradient, the oxygen permeation flux of $Ba_{0.9}Co_{0.7}Fe_{0.2}Nb_{0.1}O_{3-\delta}$ membrane was 1.8 $ml \cdot min^{-1} \cdot cm^{-2}$ for a 1 mm thick membrane at 875 °C. Further, the electrolysis cell has showed very stable performance during a 200 h short-term electrolysis testing for LSGM electrolyte supported cells with $Ba_{0.9}Co_{0.5}Fe_{0.4}Nb_{0.1}O_{3-\delta}$ as oxygen electrode. In conclusion, this work has established a solid foundation for the practical application of mix-conducting materials.

INTRODUCTION

Energy and environment are the major issues facing mankind in the 21st century. In recent years, due to the growing energy crisis and the serious environmental pollution problems caused by consumption of traditional energy sources, mixed conducting perovskite ceramic materials have attracted significant attentions in the field of chemistry, materials and physical scientists. These materials can be used in many ambitious technologies such as Oxygen Transport Membranes (OTM), Solid Oxide Fuel Cells (SOFCs) and Solid Oxide Electrolysis Cells (SOECs).

SOFC is an electrochemical conversion device that produces electricity energy and heat directly from oxidizing a fuel, giving much higher conversion efficiencies than conventional energy system.[1] SOEC is one of the promising technologies for hydrogen economy, because it can generate hydrogen through steam electrolysis with high system efficiency. It is expected that SOEC technology can enable the transformation of intermittent power sources such as solar and wind energy into "firm power", providing a very promising efficient and viable process for large-scale production of hydrogen in the future.[2] OTM shows superior oxygen theoretically infinite selectivity. In addition to producing high purity oxygen, this kind of dense ceramic membranes can also be integrated in catalytic membrane reactors such as partial oxidation of methane to syngas (POM).[3]

In SOFCs, the cathode material is known to limit the overall system performance and therefore research focuses on the development of new cathode materials with similar or even superior properties. Compared the traditional pure electronic conductors cathode materials such as $La_{0.8}Sr_{0.2}MnO_3$ (LSM), the active site is effectively extended to the entire exposed cathode surface if a mixed conductor is used as the cathode material, such as $La_{0.6}Sr_{0.4}Co_{0.2}Fe_{0.8}O_{3-\delta}$ (LSCF).[4] However, cobalt containing perovskite materials typically lack chemically stability. Meanwhile, strontium segregation was observed in LSCF with the temperature and pO_2 of cathode. For SOEC, the long-term stability remains an open issue and a number of degradation/failure mechanisms have been reported, especially for the oxygen electrode delamination due to limited O_2 formation at the electrolyte-electrode interface.[5] Therefore, it is of great significance to develop novel oxygen electrode materials with low area specific resistance (ASR) and high long-term stability for practical development and deployment of the SOEC technology for large-scale hydrogen production. Further, for the OTMs, it is difficult for preparation well-distributed dual phase membrane.[6, 7] Therefore, single phase of mixed conducting membrane is optional for researchers.

Recently, many mixed conducting materials were researched. And many methods were elaborated for optimizing new materials for high temperature electrochemical device.[8, 9] Hereon, in order to develop new promising materials for the high temperature devices, it is necessary to classify already known perovskite materials according to their properties and to identify certain tendencies. Thereby, the composition depending on structural parameters and materials properties should be considered. Structural parameters under consideration are the Goldschmidt tolerance factor which describes the structure stability of perovskites and the binding energy were calculated to evaluate the chemical stability of perovskites.[10] The critical radius and lattice free volume are applied as geometrical measures for the ionic conductivity, and the oxygen ions formation energy and migration energy were investigated by first-principles which can describes the ionic conductivity on the microscopic level.[11] Therefore, the materials design for high temperature electrochemical devices was addressed in this paper. Experimental investigations of the designed mixed-conductor perovskites materials were performed in our laboratory, and some interesting results were obtained, which is corresponding to the theoretical calculation.

MATERIAL DESIGN FOR HIGH TMEPERATURE ELECTROCHEMICAL DEVICES

Structure Stability

The structure stability of mixed conducting perovskite materials is usually defined in terms of the Goldschmidt tolerance factor t: [12]

$$t= (R_A+ R_O) / \sqrt{2}\ (R_B+ R_O) \tag{1}$$

where R_A, R_B and R_O is the effective ionic radii of the A- and B-site elements and oxygen ions, respectively. The structural stability requires that the value of t should be as close to 1 as possible.

The calculated results of tolerance factors for some mixed conducting materials are described in Fig. 2.1. Shannon's ionic radii referring to the coordination numbers 12 (A-site) and 6 (B-site) has been used, although it is known that oxygen deficiency influences the coordination number and therefore the ionic radii. From the results, we can find that the tolerance factors of several popular materials such as LSM, LSCF are close to 1. Therefore the tolerance factor should be considered first in designing a new material.

The tolerance factor of $Ba_{1-x}Sr_xCo_{0.8}Fe_{0.2}O_{3-\delta}$（BSCF）and $Ba_{1.0}Co_{0.7}Fe_{0.2}Nb_{0.1}O_{3-\delta}$（BCFN） are slightly greater than 1 due to the large ionic radius of Ba^{2+}（161 pm） at A-site compared with La^{3+} (136 pm) and Sr^{2+}(144pm). Smaller tolerance factor can be achieved theoretically by introducing A-site deficiency, which indicates higher structural stability. Fig. 2.1 shows that the tolerance factors of $Ba_{1.0}Co_{0.9-y}Fe_yNb_{0.1}O_{3-\delta}$ decrease with the increase of iron ion proportion. At the same valence state, the ionic radius of Co^{4+}(53pm) is smaller than Fe^{4+}(58.5pm), therefore, increasing the iron ion proportion can lower the tolerance factor and increase the structural stability.

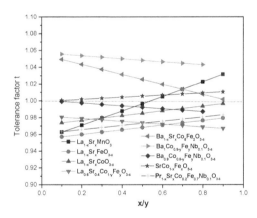

Fig.2.1 Calculated tolerance factors of the mixed conducting materials

Chemical Stability

The binding energy is referred to describe the chemical stability of perovskite materials. It is stated that the greater the binding energy is, the more stable and possibly existence the corresponding phase is.[13] Fig. 2.2 shows the binding energies in $BaBO_3$ (B = Co,Fe,Nb) estimated by first-principles calculations via the software of VASP, which is based on the density-functional theory (DFT). High valence Nb ions will increase the binding energy of $BaBO_3$-based materials. Therefore, Nb is shown to be good choices of all the present elements to improve the stability of $BaBO_3$-based materials.[14] And it is also verified by experiment results.[15] The relationship between the substitutional cations and the stability of

$Sr(Co0_{.9}X_{0.1})O_{3-\delta}$, where X is Ni, Cu, Zn, Cr, Fe, Al, Ga, In, Ce, Ti, Zr, Sn, V or Nb has been studied. Results show that the sequence of the perovskite stability upon the substituting cation is (Ni, Cu, Zn, In, Ce) < (Cr, Al, Ga, Zr, Sn, V) < Fe < Ti < Nb. Therefore, the system of $Ba_{1-x}Co_{0.9-y}Fe_yNb_{1-x}O_{3-\delta}$ was investigated in this paper due to introduction of Nb in the B-site of cobalt-based perovskites can significantly improve chemical stability.

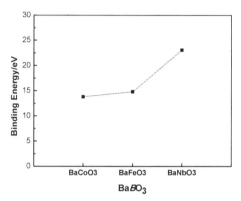

Fig.2.2 Calculated binding energies of $BaBO_3$ (B=Co,Fe,Nb) obtained with DFT calculations

Ionic Conductivity

As mentioned in the reference, the critical radius should be as large as possible in order to enhance oxygen ionic conductivity. However, a substitution of Sr^{2+} for La^{3+} decreases the critical radius, which indicates a decrease in ionic conductivity. But on the other hand, it introduces oxygen vacancies due to lower valence state in A site, which will enhance the ionic conductivity. As described for $La_{1-x}Sr_xCo_{0.8}Fe_{0.2}O_{3-\delta}$, the ionic conductivity increased monotonically for increasing x,[16] which indicates the vacancy formation to be more important than the critical radius. Therefore Ba^{2+}, Sr^{2+} et al should be doped in A site for enhance the ionic conductivity.

On another hand, the experimental results on the relation between ionic conductivity and lattice free volume were found. The ionic conductivity is increased with the lattice free volume [17]. The free volume is defined as the free space unoccupied by atoms in the lattice and is calculated using ionic radii. Fig.2.3 shows the calculated lattice free volume of perovskite materials. From the results, we can find that the ionic conductivity increases along with the radius of A-site and B-site ions. For example, higher ionic conductivity can be achieved when the A-site ions was doped by Ba, such as $Ba_{0.5}Sr_{0.5}Co_{0.8}Fe_{0.2}O_{3-\delta}$.

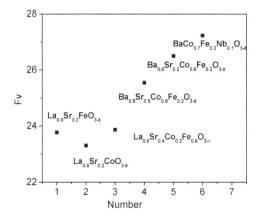

Fig.2.3 The calculated results of freedom volume

Electronic Conductivity

The oxygen ionic conductivity has been reported to be at least two orders of magnitude lower than that of the electronic conductivity in the $BaCoO_{3-\delta}$based mix-conductor perovskite materials. Consequently, the conductivity of $Ba_{0.9}Co_{0.9-y}Fe_yNb_{0.1}O_{3-\delta}$ ($0.2\leq y\leq 0.6$) shown in the references can be assumed to be primarily electronic conductivity which were reported by us before.[18] The electrical conductivity of $Ba_{0.9}Co_{0.9-y}Fe_yNb_{0.1}O_{3-\delta}$ ($0.2\leq y\leq 0.6$) decreases with the increase in Fe-doping concentration, agreeing well with the fact that the mobility of iron ions is much lower than that of cobalt ions. The same decreasing trend of electrical conductivity with the increase in iron concentration has also been observed in the system of $La_{0.8}Sr_{0.2}Co_{1-y}Fe_yO_{3-\delta}$.[19]

A Site Deficiency

It has been demonstrated that using A-site deficiency BSCF can create additional oxygen vacancies, facilitate oxygen reduction reaction and promote oxygen ion diffusion within the cathode material.[20] Consequently, A-site deficient $Ba_{1-x}Co_{0.7}Fe_{0.2}Nb_{0.1}O_{3-\delta}$ has been systematically evaluated as cathode for $La_{0.8}Sr_{0.2}Ga_{0.83}Mg_{0.17}O_{3-\delta}$ (LSGM) electrolyte supported intermediate temperature solid oxide fuel cells in our prior work.[17] When the Ba deficiency concentration increases up to 0.10, the cathode polarization resistance decreases, which will be beneficial for fuel cell performance.

For another, it is concluded that by introducing deficiency at A-site of $Ba_{0.9}Co_{0.7}Fe_{0.2}Nb_{0.1}O_{3-\delta}$, $Ba_{0.9}Co_{0.7}Fe_{0.2}Nb_{0.1}O_{3-\delta}$ can achieve a better stability than $Ba_{1.0}Co_{0.7}Fe_{0.2}Nb_{0.1}O_{3-\delta}$ in CO_2 at high temperature in our prior work.[21]

APPLICATION OF $Ba_{1-x}Co_{0.9-y}Fe_yNb_{0.1}O_{3-\delta}$ IN HIGH TEMPERATURE ELECTROCHEMICAL DEVICES

$Ba_{0.9}Co_{0.7}Fe_{0.2}Nb_{0.1}O_{3-\delta}$ for OTM

In our prior work, A-site deficient perovskite $Ba_{0.9}Co_{0.7}Fe_{0.2}Nb_{0.1}O_{3-\delta}$ was introduced in order to enhance the stability and the oxygen permeability simultaneously. Under an air/helium oxygen gradient, the oxygen permeation flux of 1 mm thick $Ba_{0.9}Co_{0.7}Fe_{0.2}Nb_{0.1}O_{3-\delta}$ membrane was 1.8 ml·min^{-1}·cm^{-2} at 875 °C, which is slightly higher than that of $BaCo_{0.7}Fe_{0.2}Nb_{0.1}O_{3-\delta}$ membrane under the same condition.[21] The BCFN membrane was further investigated for the partial oxidation of methane to syngas with NiO/MgO as the reforming catalyst. At 875°C, a permeation flux of 7.4 ml.cm^{-2}.min^{-1}, methane conversion of 92%, and CO selectivity of 94% were achieved. A 400 hours durability test of $Ba_{0.9}Co_{0.7}Fe_{0.2}Nb_{0.1}O_{3-\delta}$ membrane for POM reaction at 875 °C showed no obviously degradation in the measured oxygen permeation flux and methane conversion, which indicated that $Ba_{0.9}Co_{0.7}Fe_{0.2}Nb_{0.1}O_{3-\delta}$ could be a promising material for OTM.

Further, the dense BCFN OTM structure was optimized, and a porous-dense-porous three-layer BCFN oxygen transport membrane was devised and prepared. The experimental results show that the three-layer structure promoted the oxygen permeation flux up to 1.8 times than 0.6 mm thick dense BCFN membrane, the oxygen permeation fluxes shows no degradation and the perovskite structure keeps stable in the 100 h operation.

$Ba_{0.9}Co_{0.7}Fe_{0.2}Nb_{0.1}O_{3-\delta}$ for SOFC Cathode

$Ba_{0.9}Co_{0.7}Fe_{0.2}Nb_{0.1}O_{3-\delta}$ (BCFN) perovskite material was synthesized and evaluated as cathode for $La_{0.8}Sr_{0.2}Ga_{0.83}Mg_{0.17}O_{3-\delta}$ (LSGM) electrolyte supported intermediate temperature solid oxide fuel cells (IT-SOFCs).[22] X-ray diffraction results showed that BCFN was chemically compatible with the LSGM electrolyte. Maximum power densities of 0.36, 0.57, 0.80 and 1.1 W/cm^2 were obtained for LSGM electrolyte supported cells with BCFN as cathode and Ni-GDC as anode operated at 650, 700, 750 and 800 °C, respectively. Further, the cell performance was stable under a constant current of 0.6 A/cm^2 for over 204 h at 750 °C.

$Ba_{0.9}Co_{0.5}Fe_{0.4}Nb_{0.1}O_{3-\delta}$ for SOEC Oxygen Electrode

Electrochemical impedance spectra and voltage-current curves of the electrolysis cell with the configuration of $Ba_{0.9}Co_{0.5}Fe_{0.4}Nb_{0.1}O_{3-\delta}|LSGM|La_{0.4}Ce_{0.6}O_{2-x}|Ni-Ce_{0.9}Gd_{0.1}O_{1.95}$ were measured as a function of operating temperature and steam concentrations to characterize the electrolysis cell performances.[23] Under an applied electrolysis voltage of 1.5 V, the maximum consumed current density increased from 0.835 A cm^{-2} at 700 °C to 3.237 A cm^{-2} at 850 °C with 40 vol.% absolute humidity (AH), and the hydrogen generation rate of the cell can be up to 1352 mL cm^{-2} h^{-1} with 40 vol.% AH at 850 °C. Further, the electrolysis cell has showed very stable performance during a 200 h short-term electrolysis testing, indicating that $Ba_{0.9}Co_{0.5}Fe_{0.4}Nb_{0.1}O_{3-\delta}$ can be a very promising candidate for the oxygen electrode of SOECs using LSGM electrolyte.

CONCLUSIONS

In this study, $Ba_{1-x}Co_{0.9-y}Fe_yNb_{0.1}O_{3-\delta}$ has been introduced based on the study of calculate analysis- experimental data. We find out the optimized scope and rules of the composition and properties of the existing mix-conductor materials. The performance and stability of $Ba_{1-x}Co_{0.9-y}Fe_yNb_{0.1}O_{3-\delta}$ (x=0-0.15, y=0-0.9) for high temperature electrochemical device applications such as OTM, SOFC and SOEC have been introduced. The results show that $Ba_{0.9}Co_{0.7}Fe_{0.2}Nb_{0.1}O_{3-\delta}$ can be a very promising candidate for the cathode of SOFCs and membrane materials of OTM application. In conclusion, this dissertation has established a solid foundation for the practical application of mix-conducting materials.

ACKNOWLEDGEMENTS

Financial support from 973 project (contract no 2012CB215404) and the NSFC-NSF cooperation project of China-US (award no 51261120378) is greatly appreciated.

REFERENCES

[1] S.C Singhal, K. Kendall, High Temperature Solid Oxide Fuel Cells: Fundamentals, Design and Applications, *Elsevier Advanced Technology*, Oxford, 2003.

[2] Hauch A, Ebbesen SD, Jensen SH, Mogensen M, Highly Efficient High Temperature Electrolysis, *J Mater Chem*, 18, 2331-40 (2008).

[3] Sunarso J, Mixed Ionic-electronic Conducting (MIEC) Ceramic-based Membranes for Oxygen Separation, *Journal of Membrane Science,* 320, 13-41 (2008).

[4] Chunwen Sun, Rob Hui, Justin Roller, Cathode Materials for Solid Oxide Fuel Cells: A Review, *Journal of Solid State Electrochemistry*, 14,1125-1144 (2010).

[5] Jiang SP, Love JG, Observation of Structural Change Induced By Cathodic Polarization on $(La,Sr)MnO_3$ Electrodes of Solid Oxide Fuel Cells, *Solid State Ionics,* 158, 45-53 (2003).

[6] Chusheng Chen et al, Preparation and Oxygen Permeability of $Ce_{0.8}Sm_{0.2}O_{2-\delta}$-$La_{0.7}Ca_{0.3}CrO_{3-\delta}$ Dual-phase Composite Hollow Fiber Membrane, *Solid State Ionics,* 225, 690-694 (2012).

[7] Weishen Yang et al, Design and Experimental Investigation of Oxide Ceramic Dual-phase Membranes, *Journal of Membrane Science*, 120-130, 394-395 (2012).

[8] M. Søgaard, Transport Properties and Oxygen Stoichiometry of Mixed Ionic Electronic Conducting Perovskite-type Oxides, *Risø National Laboratory*, Roskilde, Denmark, 2006.

[9] E. Girdauskaite, H. Ullmann, M. Al Daroukh, et al., Oxygen Stoichiometry, Unit Cell Volume, and Thermodynamic Quantities of Perovskite-type Oxides, *Journal of Solid State Electrochemistry*, 11, 469-477 (2007).

[10] J. Richter, Mixed Conducting Ceramics for High Temperature Electrochemical Devices, [PhD thesis], ETH, Zurich, 2008.

[11] Xue Li, Hailei Zhao, Feng Gao, et al., La and Sc Co-doped $SrTiO_3$ as Novel Anode Materials for Solid Oxide Fuel Cells, *Electrochemistry Communications*, 10, 1567-1570 (2008).

[12] T. Ishihara et al., Perovskite Oxide for Solid Oxide Fuel Cells, *Springer*, 2009.

[13] Li-an Yu, Lu Jin, Min-fang Han,First-principles Study on Electronic Structures and Oxygen Vacancy Energies of Perovskite-type Oxides $BaBO3-\delta$(B = Fe,Co,Nb),*Journal of Synthetic Crystals*, 41(3),747-752(2012).

[14] Na Li, Ning Chen, Fushen Li, et al., Theoretical Research on Optimization Ingredient Regulation of $BaBO_3$ series hypoxic materials, *Scientia Sinica Phys, Mech & Astron*, 41(9), 1075-1079 (2011).

[15] T. Nagai, W. Ito, T. Sakon, Relationship Between Cation Substitution And Stability of Perovskite Structure in $SrCoO_{3-\delta}$-based Mixed Conductors, *Solid State Ionics*, 177, 3433-3444 (2007).

[16] Teraoka Y, Nobunaga T, Okamoto K, Miura N, Yamazoe N, Influence of Constituent Metal Cations in Substituted $LaCoO_3$ on Mixed Conductivity And Oxygen Permeability, *Solid State Ionics*, 48, 207-212 (1991).

[17] Jianhua Tong, Weishen Yang. Methods for Selection Perovskite-type Oxides Used as

Mixed-conducting Membranes for Oxygen Separation, *Membrane science and technology*, 23, 33-42 (2003).

[18]Zhibin Yang, Minfang Han, Peiyu Zhu, et al., $Ba_{1-x}Co_{0.9-y}Fe_yNb_{0.1}O_{3-\delta}$(x = 0-0.15, y = 0-0.9) as Cathode Materials for Solid Oxide Fuel Cells, *International Journal of Hydrogen Energy*, 36, 9162-9168 (2011).

[19]Tai LW, Nasrallah MM, Anderson HU, Sparlin DM, Sehlin SR, Structure and Electrical Properties of $La_{1-x}Sr_xCo_{1-y}Fe_yO_{3-\delta}$, Part 2, The System $La_{0.8}Sr_{0.2}Co_{1-y}Fe_yO_{3-\delta}$, *Solid State Ionics*, 75, 273-83 (1995).

[20]W. Zhou, R. Ran, Z.P. Shao, W.Q. Jin, N.P. Xu, Evaluation of A-site Cation-deficient $(Ba_{0.5}Sr_{0.5})_{1-x}Co_{0.8}Fe_{0.2}O_{3-\delta}$ (x > 0) Perovskite as A Solid Oxide Fuel Cell Cathode, *J. Power Sources*, 182, 24-31 (2008).

[21]Shidong Song, Peng Zhang, Minfang Han,et al., Oxygen Permeation and Partial Oxidation of Methane Reaction in $Ba_{0.9}Co_{0.7}Fe_{0.2}Nb_{0.1}O_{3-\delta}$ Oxygen Permeation Membrane, *Journal of membrane science*, 415-416, 654-662 (2012).

[22]Zhibin Yang, Chenghao Yang, Chao Jin, et al., $Ba_{0.9}Co_{0.7}Fe_{0.2}Nb_{0.1}O_{3-\delta}$ as Cathode Material for Intermediate Temperature Solid Oxide Fuel Cells, *Electrochemistry Communications*, 13, 882-885 (2011).

[23]Zhibin Yang, Chao Jin, Chenghao Yang, et al., $Ba_{0.9}Co_{0.5}Fe_{0.4}Nb_{0.1}O_{3-\delta}$ as Novel Oxygen Electrode for Solid Oxide Electrolysis Cells. *International Journal of Hydrogen Energy*, 36, 11572-11577 (2011).

IONIC CONDUCTIVITY IN MULLITE AND MULLITE TYPE COMPOUNDS

C. H. Rüscher, F. Kiesel
Leibniz Universität Hannover, Institut für Mineralogie,
Callinstr. 3, 30167 Hannover, Germany
E-Mail: claus.ruescher@mineralogie.uni-hannover.de

ABSTRACT

Mullite, $Al_{4.8}Si_{1.2}O_{9.6}$, and mullite type componds $Bi_2M_4O_9$ with M = Fe, Al were investigated by impedance spectroscopy between 1 Hz and 1 MHz and temperatures between 200°C and 800°C. The data reveal universal dielectric response (UDR) type behaviour which could be separated as $\sigma(\nu) = \sigma_0 + A\nu^s$ and $\varepsilon(\nu) = \varepsilon_{st} + A\nu^{s-1}\tan(s\pi/2)/\varepsilon_0$, however, with the s parameter dependent on temperature and frequency. Specific σ_0 values of the crystal plates could be extracted which are about 10^{-8} S/cm for mullite, 10^{-7} S/cm for $Bi_2Al_4O_9$ and 10^{-6} S/cm for $Bi_2Fe_4O_9$ taken at 500°C. The main peak in the Modulus representation M'' is given by $2\pi\nu_c = \sigma_0/\varepsilon_{st}\varepsilon_0$ identifying the net rate of thermally activated excitations $2\pi\nu_c = 2\pi\nu_0\exp(-E_a/kT)$ of relevant charge carriers. Ea of about 1.3 eV and 1 eV are obtained for $Bi_2M_4O_9$ and mullite, respectively. With increasing temperature effects related to ion blocking at electrodes are observed towards lower frequencies. It is discussed that for mullite σ_0 is related to impurity charge carriers (Na^+, H^+, etc.) whereas for $Bi_2Al_4O_9$ and $Bi_2Fe_4O_9$ "intrinsic" diffusion of Al, Fe via Frenkel type defects may not be ruled out.

INTRODUCTION

Impedance spectroscopy is an important tool for investigations of ionic conduction. The absolute magnitude and related interpretations of existing impedance data for ceramics of mullite and mullite type compounds are, however, rather controversial. For mullite type $Bi_2Al_4O_9$ ceramic samples Bloom et al.[1] observed an extraordinary high conductivity, about 10^{-2} S/cm at 800°C which has been related to diffusion of oxygen. Zha et al.[2] reported a further increase in conductivity (up to 0.3 S/cm at 800°C) in the system related on doping $Bi_{2-2x}Sr_{2x}Al_4O_{9-x}$. This could be understood with the formation of oxygen vacancies which could be controlled by the doping effect. Therefore, these materials could be considered for future SOFC (solid oxide fuel cell) applications. Contrary to these optimistic reports Larose and Akbar[3] concluded that $Bi_2Al_4O_9$ could not be a fast oxygen ion conductor and explained the observed conductivity by a combined effect of oxygen ions and electrons. But any significant contribution of oxygen ions to the conductivity should be negligible due to the rather small oxygen diffusion coefficients extracted from $^{18}O/^{16}O$ exchange experiments D_O between 10^{-15} and 10^{-14} cm²/s at 800°C.[4-7] Finally it has been shown that rather high conductivities in mullite type compounds could be realized by Bi_2O_3 additions on grain boundaries[8]. Explanations for the bulk conductivities remained open. For mullite ceramics conductivity values as extracted from impedance spectra were reported to be about $1.5*10^{-5}$ S/cm at 1400°C and $7*10^{-9}$ S/cm at 500°C suggesting an activated behavior with E_a about 1 eV. This has been taken as an indication of an appropriate gap for some electronic excitation. However, mullite is an insulator showing a much larger bandgap far in the UV making any electronic contribution to the conductivity negligible. Therefore, impedance spectra of mullite and mullite type compounds were reinvestigated comparing single crystal data and data obtained on ceramic samples.

EXPERIMENTAL

Sample preparation and measurement technique

Polycrystalline material of $Bi_2M_4O_9$ with M = Fe, Al were prepared following the route described earlier.[7] Herein metal nitrates were taken in the required stoichiometric ratio and were mixed in glycerine. The liquid was slowly heated to 80°C obtaining a gel, which were subsequently heated at 120°C for about 12 h, followed by calcinations at 800°C for 24 h. The samples then were washed in acidic nitrate which removes secondary phases. Characterisations of these samples by XRD and IR-absorption shows the rather pure and well crystallized properties in very good agreement with the results reported [6] $Bi_2Fe_4O_9$ could be pressed and successively sintered at 800°C into pellets (13 mm diameter, 1 mm thick) of suitable mechanical stability and density of up to about 90% for further investigations. To obtain pellets of similar mechanical quality of $Bi_2Al_4O_9$ samples PVA (C_2H_4O) were added to the as prepared polycrystalline sample followed by pressing and sintering. The density could, however, not be better than 70%.

For $Bi_2Fe_4O_9$ a crystal plate of 5*4 mm^2 and 1 mm thickness could be used for the investigation. The orientation of the crystal plate could be checked by polarized IR reflection spectroscopy to be (010). The crystal piece of $Bi_2Al_4O_9$ had a thickness of 1 mm and just enough area to fix the electrodes for the measurements. The orientation could also be determined as for $Bi_2Fe_4O_9$ showing that the component E parallel to c was measured in the impedance spectra.

Oriented fully transparent mullite crystal plates of sizes of about 3*3 mm^2 and 1 mm thickness were used. The origin of crystals, growth conditions and quality, are given earlier [5]. It has been checked that the composition is well $Al_{4.8}Si_{1.2}O_{9.6}$ (2:1 mullite). Impurity concentrations (H_2O, Na_2O) are about 10 ppm [9].

Impedance measurements were carried out between 1 Hz and 1 MHz on pieces of the ceramics covered with sputtered Au discs of 3 mm diameter on opposite side. In the same way the $Bi_2Fe_4O_9$ crystal plate was Au sputtered. The crystal piece of $Bi_2Al_4O_9$ was taken without sputtering. For mullite special Ag-paste related electrode were used, which were dried at 500°C before use. The samples were fixed between two Pt electrodes and carried into the heating device (Nabertherm) using an in-house constructed arrangement for the sample holder and connecting wires. Herein each electrode is connected via two separated and shielded wires to the impedance spectrometer for the applied current supply and the voltage measurement, respectively. During the measurements the sample was shielded, too. The measurement temperatures range between 200 and 800°C. Heating device and impedance spectrometer (Alpha-A High Performance Frequency Analyzer) were run computer controlled for the measurements.

Data handling and equations used for evaluation

Nyquist diagrams based on the measured impedance Z(v) = R + iX (R, Z = real and imaginary part of the impedance, v = frequency in Hz) were considered. The data were converted into frequency dependent conductivity ($\sigma(v)$ = d/F *G) and relative permittivity ($\varepsilon(v)$ = d/F*B/ω* 1/ε_0, ω being the angular frequency, ω = $2\pi v$, and ε_0 the Faraday constant ($8.85*10^{-14}$ F/cm)). B and G are given by Y(v) = 1/Z = G + iB. F and d are area of electrode F and thickness of sample, respectively.

The universal dielectric response (UDR) function [10, 11]

$$\sigma(\omega) = \sigma_{dc} + A'\omega^s \tag{1}$$

was used including the zero frequency parameter σ_0. Generally the spectra $\sigma(\omega)$ and $\varepsilon(\omega)$ are related by the Kramers-Kronig relation (KK).[10, 11] Therefore the real part of permittivity reads [12]

$$\varepsilon(\omega) = \varepsilon_{st} + A'\omega^{s-1}\tan(s\pi/2)/\varepsilon_0 \tag{2}$$

In eq. 2 ε_{st} is added to realize the quasi static contribution of permittivity due to polarizability of the lattice plus electronic part. Additional dipole contribution with relaxations between frequencies as realized here and the far infrared could contribute. A separation of the frequency independent and frequency dependent contributions could be obtained using first derivatives of eq. 1 and 2, namely

$$d\sigma(\omega)/d\omega = A's\omega^{s-1} \tag{3}$$

and

$$d\varepsilon(\omega)/d(\omega) = A'(s-1)\omega^{s-2}\tan(s\pi/2)/\varepsilon_0 \tag{4}$$

Eqs. 3 and 4 become straight lines in log-log plots combining slope (s-1) and intercept (log(A'*s)) in conductivity and slope (s-2) and intercept log(A'(1-s)tan(sπ/2)/ε_0) in permittivity.
   The Debye function

$$\varepsilon(\omega) = S_{Db}/(1+i\omega/\omega_{Db}) \tag{5}$$

was used and added adequately to equations 1 and 2, i.e a Debye relaxation centered at ω_{Db} of strength S_{Db}.
   For better comparison the "electric" modulus representation as introduced by Macedo et al. [13] as an analytical scheme to extract long range ionic conduction processes in impedance spectra were used as:

$$M^* = M' + iM'' = 1/\varepsilon^* = \varepsilon'/(\varepsilon'^2 + \varepsilon''^2) + i\varepsilon''/(\varepsilon'^2+\varepsilon''^2) \tag{6}$$

Herein a peak in M'' and a ramp in M' is centered at the inverse conductivity relaxation time

$$\omega_c = 1/\tau_c = \sigma_{dc}/\varepsilon_{st}\varepsilon_0 \tag{7}$$

Since the spectra are given as a function of the measuring frequency the functions used were reformulated with $\omega = 2\pi\nu$, eg. revealing parameters A= A'$(2\pi)^s$, $\nu_{Db} = \omega_{Db}/2\pi$, $\nu_c= \omega_c/2\pi$.

RESULTS

Impedance spectra: general features
   Spectra are shown for selected temperatures for the logarithm of the real part of conductivity (Fig. 1 a) and the real part of permittivity (Fig. 1 b) as a function of logarithm of frequency (log(v)) for $Bi_2Fe_4O_9$ crystal plate. The spectra shown were taken in the cooling down run. Spectra for the cooling and heating run coincide above about 400°C. With increasing temperature some burning effect in the electrodes and strengthening of the contact pressure could be the reason for deviations in the spectra from those of the cooling down run below 400°C. At the lowest temperature shown a rather flat behaviour is observed in $\varepsilon(v)$ whereas the conductivity spectra show a steep increase with increasing temperature. With increasing temperature a flat behaviour becomes gradually more dominant in $\sigma(v)$ and the strong frequency dependence shifts gradually to higher frequencies. Contrary the permittivity becomes gradually more frequency dependent in the low frequency region. Similar behavior was also observed for the $Bi_2Fe_4O_9$ ceramic plate. Qualitatively similar behavior was also obtained for $Bi_2Al_4O_9$ and mullite crystal plates. How-

ever, the crossover to dominating flat behaviour in σ(v) could be seen at significant higher temperature for $Bi_2Al_4O_9$ above about 450°C and for mullite above about 550°C.

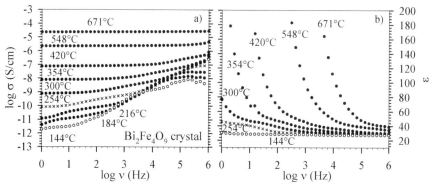

Figure 1. Spectra of real part of a) conductivity σ(v) and b) permittivity ε(v) for $Bi_2Fe_4O_9$ for selected temperatures.

Impedance spectra: data evaluation

All spectra could be further evaluated using diagrams as shown for the $Bi_2Fe_4O_9$ crystal plate in Fig. 2 a-g at 144°C and 3 a-g at 385°C. Shown are various representations, which could help to separate the different effects. Unlike all other diagrams where geometrical correction are included, the Nyquist diagrams (2 a, 3 a) show the as received data, i.e. without geometrical correction. The data show at the lower temperature a small part of a semicircle. With increasing temperature the semicircle becomes more complete and a tailing evolves in the low frequency range. The main semicircle is related to the RC-type bulk property of $Bi_2Fe_4O_9$, whereas the tailing is related to the "electrode effect", i.e the blocking of the relevant ionic charge carriers. In the permittivity (Fig. 3d) the blocking effect leads to a strong increase in values with decreasing frequency below about 100 Hz. This contribution is not further considered in the calculations as could be seen by the solid and dashes lines in Fig. 3. The solid lines result from calculations using eq. 1 and 2 for the appropriate representations. The influence on the obtained curve due to the Av^s behavior is almost negligible in the Nyquist diagrams. Similarly the effect of a Debye type contribution (eq. 5) as included in the dashed line is almost negligible in the appearance of the Nyquist diagrams, too. Thus from the Nyquist diagram alone only the main value of the frequency independent contribution of conductivity can be obtained by extrapolating the semicircle down to the zero value in Re(Z). The frequency at the maximum value in the semicircle corresponds to the relaxation frequency of an equivalent RC cirquit. Following Macedo et al. [13] it is related to the "conductivity relaxation time" (eq. 7) and determines the rate of decay of the electrical field. It necessarily appears as the peak maximum in the modulus function M'' (Fig. 3 f) and the center of the main ramp in M' (Fig. 3 g). Additionally to the conducting effect in the dielectric medium another feature described closely by a Debye formalism is seen in the Modulus representation. The separation of this contribution is also seen by inspecting the conductivity spectra (Fig. 3 b) and even more clearly in their derivatives (Fig. 3 c) following the calculated solid (Debye included) and dashed curves (Debye neglected). Since the derivatives are necessarily free from the quasi static contributions (eq. 1, 3: $σ_0$; eq. 2, 4: $ε_{st}$), this data treatment shows all frequency dependent contributions.

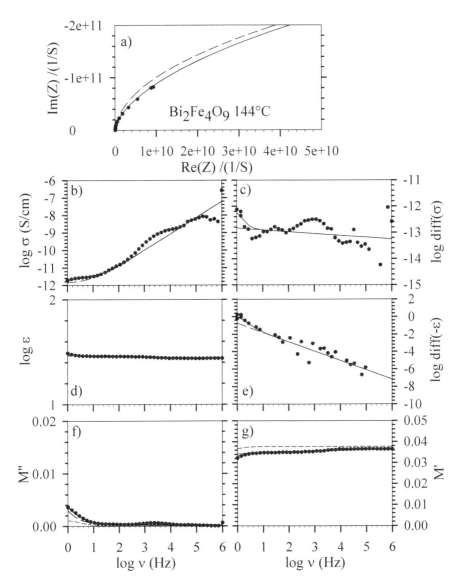

Figure 2. Impedance spectra (closed circles) of $Bi_2Fe_4O_9$ crystal sample at 144°C in various presentations: a) Nyquist diagram, and as a function of log(ν) for b) log(σ), c) log(Δσ/Δν), d) log(ε), e) log(Δε/Δν), f) imaginary part of modulus function M'', g) real part of modulus function M'. Solid and dashed lines are calculated following eq. 1-6. For details see text.

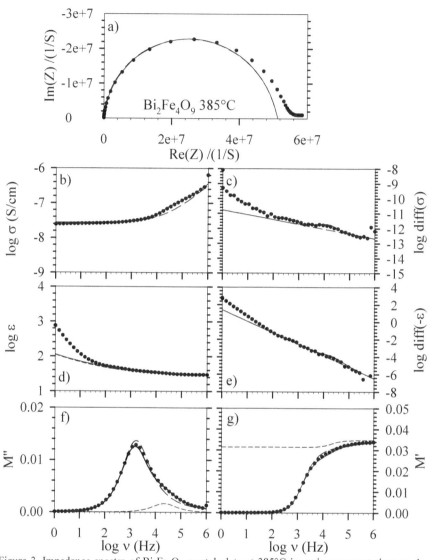

Figure 3. Impedance spectra of $Bi_2Fe_4O_9$ crystal plate at 385°C in various presentations as described in Fig. 2. Details are given in the text.

At the lower temperature the conductivity is dominated by $A\omega^s$ behaviour (Fig. 2 b). With s close to 1 the derivative becomes almost flat (Fig. 2 c). Debye type effects at around 1.5 kHz and below 10 Hz also contribute. Only the later has been considered tentatively in the calculation. It is interesting to note that the permittivity shows flat behaviour but the expected frequency dependence according to the KK relation is clearly seen in the first derivative spectrum (Fig. 2d, e).

Direct inspection of Fig. 3 c and e reveal s approximately 2/3 at frequencies between 100 Hz and 1 MHz, i.e s-2 = -4/3 (Fig. 3e) and s-1 = -1/3 (Fig. 3c). However, for measurements to higher frequencies an increase in s could be expected. This may also be concluded following the behaviour towards lower temperatures, observing an increase in s values (compare also Fig. 1 a). Therefore, s seems to crossover from values close to one to s => 0 only due to the limited frequency range investigated. A real s = 0 dependence may hardly be seen, it should result into a frequency independent plateau in the first derivative. It may be indicated in Fig. 3 c and e only in the limited frequency range between 100 Hz and 1 kHz. Otherwise s seems to approximate zero due to blocking effect of the electrode towards lower frequencies.

Temperature dependence of extracted parameters: relative effective permittivity, effective conductivity and inverse conductivity relaxation time

The values of the relative permittivity of mullite (E//c) and mullite type compounds $Bi_2Al_4O_9$ (E//c) and $Bi_2Fe_4O_9$ (E//b) as taken from the measured spectra at 1 MHz are shown as a function of temperature in Fig. 4. For mullite ceramic samples values of about 6-7 were notices by Gerhardt and Ruh [14] for frequencies between 100 Hz and 10 MHz. Using infrared reflectivity data and obtained phonon oscillator strength by Rüscher [15] reveal components $\varepsilon_a \approx 6.7$ $\varepsilon_b \approx 6.7$ $\varepsilon_c \approx 6.7$ for the "static permittivity", which includes all lattice and electronic contributions. The good agreement with values obtained by Gerhardt and Ruh [14] imply that no significant further contributions, e.g. from dipole orientations, could be seen in the frequency range in between. Our present values overestimate the true relative permittivity probably due to uncertainties in the geometrical factor or other electrode problems by a factor of more than 3. Electrode problems and rather uncontrolled influences of space charges due to the low density of ceramic type pellets could also strongly effect the data obtained on $Bi_2Al_4O_9$ (compare also below for the conductivity, Fig. 5). However the relative values depending on changes on temperature and frequency are certainly accurate within 5%. For $Bi_2Fe_4O_9$ values obtained on ceramics and crystal plate are well at the same level, too.

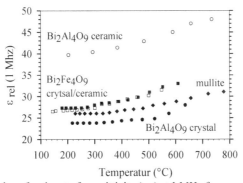

Figure 4. Extracted value of real part of permittivity (ε_{rel}) at 1 MHz for compositions as denoted.

The temperature dependence in the relative permittivity shows basically a linear increase. A more significant increase in $\varepsilon(1MHz)$ as observed for mullite above about 300°C, for $Bi_2Al_4O_9$ above 500°C and for $Bi_2Fe_4O_9$ above about 250°C which is related to temperature dependencies of in-coming Debye type contributions due to ionic relaxations. The temperature dependencies of this type of relaxations follow almost the thermally activated behaviour as observed for σ_{dc}. It is assumed that such types of relaxations are due to ionic redistributions that occur in advance or follow successful excitations for σ_{dc}.

In Fig. 5 Arrhenius plot of the extracted σ_{dc} values of the mullite type compounds are shown in comparison to values reported for $Bi_2Al_4O_9$[3]. The activation energies obtained for the compositions in this study by linear regressions are given in the Figure including the data points shown. For $Bi_2Fe_4O_9$ single crystal and ceramic data well coincide. This may not easily imply that influences of grain boundary effects may be negligible, but that the $Bi_2Fe_4O_9$ crystal sample contains significant intergrowth phenomena. On the other hand the significant difference in dc conductivity values between single crystal and ceramic data for $Bi_2Al_4O_9$ are certainly related to the much lesser density obtained for the later samples probably together with a grain boundary contribution. Larose and Akbar[3] could well separate a grain boundary arc and bulk related arc in the Nyquist diagrams obtaining significantly higher dc values for the bulk. The single crystal data obtained in this study show still higher values and also a significant higher activation energy compared to that obtained by Larose and Akbar[3].

The frequency of the peak maximum in modulus M'' is shown in Fig. 6 in an Arrhenius type plot as obtained for the ceramics and crystal samples. Linear behaviour could be observed with activation energies obtained by linear regressions to the data as shown in the figure. Also shown by crosses are values calculated using experimental data for ε_{rel} (Fig. 4) and σ_{dc} (Fig. 5) with eq. 7. It can be seen that the conduction frequency ν_c is almost perfectly determined by the frequency independent effective conductivity and the effective relative "static" permittivity, i.e. eq. 7. For $Bi_2Al_4O_9$ ceramic sample the data given certainly underestimate systematically the expected bulk values in the same way as the conductivity does. For $Bi_2Fe_4O_9$ the values closely coincide for the ceramic sample and the crystal. This could imply that influences of grain boundary effects could also be significant in the $Bi_2Fe_4O_9$ crystal sample similar to the ceramic sample as suggested above. According to this somewhat higher values of σ_{dc} and ν_c could be expected for a more perfect $Bi_2Fe_4O_9$ crystal sample.

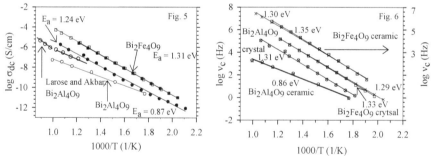

Figure 5. (left) Arrhenius plot of extracted σ_{dc} for compositions as denoted with activation energies as obtained by linear regressions (solid lines). Data reported by Larose and Akbar[3] for $Bi_2Al_4O_9$ are given for comparison as discussed in the text.

Figure 6. (right) Arrhenius plot of extracted peak position in M'' for compositions as denoted and given by open symbols. Crosses mark calculated data $2\pi\nu_c = \sigma_{dc}/\varepsilon_0\varepsilon_{st}$. Solid lines show results of linear regressions with activation energies as labeled.

For mullite Arrhenius plots of the extracted zero frequency conductivity data σ_{dc}, $T^*\sigma_{dc}$, and ν_c shown in Fig. 7. For ν_c open circles denote values directly optained from modulus M'' peak maximum and crosses were calculated using eq. 7. Again it can be seen that both set of data coincide. The activation energies obtained by linear regressions are given in the Figure, ignoring a possible change in activation energie due systematic deviations from linear behaviour of the data. Also given are data reported by Chaudhuri et al. [16] for mullite ceramic samples indicating some agreement in absolute values and also in "average activation energy" for conductivity. Linear extrapolation of the σ_{dc} values of mullite down to room temperature point to specific resistivities well above about 10^{14} Ωcm.

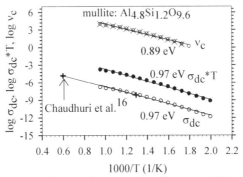

Figure 7. Arrhenius plots of σ_{dc}, $T*\sigma_{dc}$ and v_c value of mullite crystal. Values reported by Chaudhuri et al [16] for 3:2 mullite ceramics for σ_{dc} are also shown.

DISCUSSION AND CONCLUSIONS

Charge transport in mullite

The rather low activation energy of about 1 eV and low conductivity shows that any electronic contribution to the conductivity can be ruled out since the optical gap should be far in the UV spectral range, i.e. about 7 eV [17]. Thus the data may be compared to some existing ion diffusion data of mullite as shown in Fig. 8. ^{18}O, ^{26}Al and ^{30}Si tracer diffusion show thermally activated behaviour with activation energies between 3 and 5 eV given for temperatures between 1200 and 1500°C [18]. Suggesting even strongest changes in the mechanism with decreasing temperature (decrease in activation energies into the extrinsic range of diffusion) any direct contribution of the main components to the electrical transport can be ruled out. Impurity diffusion in mullite show by about 8 orders of magnitude higher values with activation energy of about 2 eV parallel to the c direction. High concentrations of impurities at the surface even lead to melting effects and enhanced inward diffusion. For mullite crystals containing about 10 ppm (wt) H_2O in form of OH-bonds diffusion coefficients as given in Fig. 8 could be obtained by dehydration profiles at 1300, 1400 and 1500°C. For mullite crystals containing less than 1 ppm H_2O an inward diffusion for OH-formation could be observed at 1670°C under wet atmospheric conditions [19]. Eils [20] could also show, that under such conditions Na and Mg could even be incorporated. From the profile analysis a diffusion coefficient in the range 10^{-8} cm^2/s could be estimated for Na incorporation at 1670°C. For comparison the conductivity data σ_{dc} were converted into the conductivity diffusion coefficient D_c via the Nernst-Einstein equation [21]

$$D_c = kT\sigma_{dc}/Ne^2 \qquad (8)$$

with N the concentration of charge carriers and e the charge of an electron (Fig. 8). For N an impurity concentration of $N \approx 10^{-17}$ cm^{-3} were assumed related to 10 ppm Na$^+$ or H$^+$ in the crystal. It may be seen that D_c largely overestimates the impurity diffusion coefficients. But much higher impurity concentration can be ruled out. Therefore, this indicates much bigger effective transport length compared to the average unit cell parameter ($<a> = a+b+c)/3$) for σ_{dc}. For clarity of this argument the Einstein-Smoulochowski relation [21] may be used for calculating the diffusion coefficient:

$$D_h = yl^2W/6 \qquad (9)$$

(l = hopping distance, W = hopping frequency, y = geometrical factor) . Assuming W = ω_c, l = $<a>$, and y = 1 the obtained D_h may well fall into a reasonable range (Fig. 8), but cannot explain the conductivity by orders of magnitude, i.e. $D_h \ll D_c$. This implies that the "mean free path" for Na$^+$ or H$^+$ must be very long for the effective charge transport in mullite or W $\ll \omega_c$.

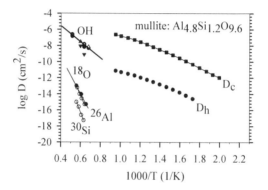

Figure 8. Arrhenius plot of tracer diffusion data of ^{18}O, ^{16}Al and ^{30}Si taken from Fielitz et al. [18], and diffusion data extracted from dehydration and hydration from Rüscher et al. [9] for mullite single crystals (2:1). D_c and D_h are diffusion coefficients obtained for mullite calculated as given in the text.

Charge transport in mullite type compounds

For Bi$_2$Al$_4$O$_9$ Larose et al. [3] suggested that the conduction mechanism could be related to oxygen diffusion with the addition of electron contribution. However tracer diffusion data obtained at 800°C reveal values between 10^{-14} and 10^{-18} cm^2/s, i.e. it seems unlikely to have significant contribution of oxygen ions to the conductivity [4-7]. On the other hand we also may rule out any significant contribution of thermally activated electrons below 1000°C due to the reasonable high optical gap. The crystal appears lightly yellow. It could be shown that there appears some tailing of the optical absorption into the blue range of light as could be measured on powders, too. For Bi$_2$Fe$_4$O$_9$ thin sections the optical absorption spectrum shows low intensity Fe^{3+} d-d transitions peaked at 1 eV and around 1.5 eV and a strong increase in intensity above 2.5 eV attributed to strong oxygen to iron charge transfer transitions. Therefore, here electronic contributions may not be ruled out. However, due to the typically very small band-width of the empty Fe-d-dominated conduction band, any excited electron may become strongly bound. Thus ruling out any significant electronic contribution and any significant oxygen contribution, the observed conductivity could still be too high to be explained by impurity conduction. Using for example equation 8 and 9 again with $W = \omega_c$, $l = <a>$, and $y = 1$ a carrier concentration n is obtained as shown in Fig. 9. The data indicate a thermal activation of carrier concentration

$$N = N_0\exp(-E_n/kT) \tag{10}$$

with E_n between 0.03 and 0.06 eV, i.e. in the range of phonon energies. The total number of carriers N_0 for $1/T = 0$ becomes close to the number of unit cells, $2.3*10^{21}$ cm^{-3}. Since the oxygen lattice seems to be rather stable this could indicate an easy formation of Frenkel type defects in the mullite type structure. For a support of this idea further investigations are required, e.g. Al/Fe exchange experiments or tracer diffusion experiments, in order to determine their diffusion coefficients.

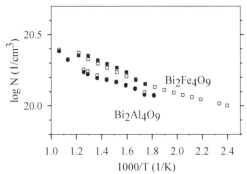

Figure 9. Calculated charge carrier concentrations N for crystal data based on the Nernst Einstein equation (eq. 8) with (open symbols) $D_c = D = a^2 2\pi v_c$ and (closed symbols) $Dc = D = a^2\sigma_{dc}/\varepsilon_{st}\varepsilon_0$ (using eq. 7), a = average lattice parameter, assuming a Haven Ratio = 1 [21].

ACKNOWLEDGEMENT

The financial support by DFG within the project GE1981/2-1 within the joint project PAK279 initiated by H. Schneider and R. X. Fischer is gratefully acknowledged. Special thanks are given to M. Mühlberg and M. Burianek who could provide crystals within the joint project and S. Ross for support in the impedance measurements.

REFERENCES
[1] I. Bloom, M. C. Hash, J. P. Zebrowski, K. M. Myles, M. Krumpelt, Solid State Ionics 53-56 (1992) 739-747.
[2] S. Zha, J. Cheng, Y. Liu, X. Liu, G. Meng, Solid State Ionics 156 (2003) 197-201.
[3] S. Larose, S. A. Akbar, J. Sol. St. Electrochem. 10 (2006) 488-498.
[4] T. Debnath, C. H. Rüscher, P. Fielitz, S. Ohmann, G. Borchardt, Ceramic Transactions 217 (2009) 71-78.
[5] T. Debnath, C. H. Rüscher, Th. M. Gesing, P. Fielitz, S. Ohmann, G. Borchardt, Ceramic Engineering and Science Proceedings 31 (2010) 81-89.
[6] T. Debnath, C. H. Rüscher, P. Fielitz, S. Ohmann, G. Borchardt, J. Sol. St. Chem. 183 (2010) 2582-2588.
[7] C. H. Rüscher, T. Debnath, P. Fielitz, S. Ohmann, G. Borchardt, diffusion fundamentals 12 (2010) 50-51.
[8] S. Ohmann, P. Fielitz, L. Dörrer, G. Borchardt, Th. M. Gesing, R. X. Fischer, C. H. Rüscher,

J. C. Buhl, H. Schneider, Sol. St. Ionics 211 (2012) 46-50.

[9]C. H. Rüscher, N. Eils, L. Robben, H. Schneider, J. Europ. Cer. Soc. 28 (2008) 393-400.

[10]A. K. Jonscher, Nature 267 (1977) 673- 679.

[11]K. L. Ngai, A. K. Jonscher, C. T. White, Nature 277 (1979) 185-189.

[12]A. Pinenov, J. Ullrich, P. Lunkenheimer, A. Loidl, C. H. Rüscher, Sol. St. Ionics 109 (1998) 111-118.

[13]P. B. Macedo, C. T. Moynihan, R. Bose, Physics and Chemistry of Glases 13 (1972) 171-179.

[14]R. A. Gerhardt, R. Ruh, J. Am. Ceram. Soc., 84 (2001) 2328-34.

[15]C. H. Rüscher, Phys. Chem. Min. 23 (1996) 50-55.

[16]S. P. Chaudhuri, S. K. Patra, A. K. Chakraborty, J. Europ. Cer. Soc., 19 (1999) 2941-2950.

[17]H. Saalfeld, W. Guse, Ceramic Transactions 6 (1990) 73-101.

[18]P. Fielitz, G. Borchardt, M. Schmücker, H. Schneider, Phil. Mag. 87 (2007) 111-127.

[19]N. Eils, C. H. Rüscher, S. Shimada, M. Schmücker, H. Schneider, J. Am. Ceram. Soc., 89 (2006) 2887-2894.

[20]N. Eils, PHD Thesi, Leibniz Universität Hannover, 2007, 1-124.

[21]G. E. Murch, Phil. Mag. A 45 (1982) 685-692.

PROTECTIVE OXIDE COATINGS FOR THE HIGH TEMPERATURE PROTECTION OF METALLIC SOFC COMPONENTS

Neil J. Kidner, Sergio Ibanez, Kellie Chenault, Kari Smith, and Matthew. M. Seabaugh
NexTech Materials
Lewis Center, OH USA

ABSTRACT

Chromia-forming, ferritic stainless steels are a leading metallic interconnect candidate due to their protective chromia scale, thermal expansion compatibility with other stack components and low cost. To achieve the required service lifetime low-cost protective coatings are necessary to provide the required oxidation resistance and chromium volatilization resistance. In this work, the viability of a low-cost aerosol-spray deposition (ASD) process for applying protective coatings to metallic interconnects is demonstrated. The oxidation resistance, electrical stability, and chromium volatilization resistance of manganese cobalt $(Mn,Co)_3O_4$ spinel protective coatings (MCO) applied by ASD to AL 441HP[TM] ferritic stainless steel has been investigated.

Excellent long-term stability (greater than 15,000 hours on test incorporating over 200 thermal cycles) of MCO coated samples has been demonstrated through electrical testing under SOFC operating conditions. In parallel, isothermal oxidation kinetics experiments have quantified a significant improvement in oxidation resistance for MCO coated versus uncoated AL 441HP[TM]. Acceleration factors based on parabolic rate constants in combination with accelerated electrical testing at 900 °C enabled an oxidation derived service lifetime for MCO coated components to be predicted. This result indicates that oxidation driven failure will not prevent MCO coated metallic interconnects from achieving their lifetime targets.

Microstructural characterization of samples at different stages of long-term testing reveal that the MCO coating is both stable under SOFC operating conditions and effective at retarding chromium diffusion and constraining chromium in the native scale at the coating/substrate interface.

INTRODUCTION

The adoption of oxidation resistant, high-temperature alloys as alternatives to traditional ceramic interconnect materials for intermediate-temperature solid oxide fuel cells (IT-SOFC) is critical to satisfying the aggressive cost targets necessary for their successful commercialization. Chromia-forming, ferritic stainless steels are a leading candidate for metallic interconnects due to their protective, and electrically conductive scale, thermal expansion compatibility with other stack components, ease of fabrication and low cost [1-2]. Engineered ferritic alloy formulations, such as Crofer® 22APU[3,4] and 22H, ZMG232L[5], and AL 441HP[TM] have been developed for SOFC metallic interconnects. Enhanced high-temperature oxidation resistance in these alloys is derived through the formation of dual-phase native scales, however, to achieve the required lifetime performance targets, protective coatings are necessary to further increase oxidation resistance and reduce chromium volatilization [6,7].

Metallic interconnects are composed of active areas (both cathode and anode), and peripheral non-active, seal areas. Both areas require excellent oxidation and chromium volatilization resistance, however, the active area also needs to be electrically conductive to minimize ohmic losses through the cell, whereas the primary function of the non-active sealant area is to be chemically inert and provide a sealing surface.

NexTech Materials (NexTech) has developed a low-cost, aerosol-spray deposition (ASD) process for applying protective oxide coatings to metallic interconnects. The process is flexible in coating composition and through simple masking allows multiple coatings to be selectively co-deposited. This versatility has enabled NexTech to pursue a coating strategy where the specific requirements of an individual area of the interconnect (cathode/anode active, non-active) can be satisfied by the selective deposited of functionally-tailored coatings. The key operations in the ASD process are outlined in Figure 1.

Figure 1. Process flow for ASD based coating process.

For the cathode active area a conductive oxide coating, such as manganese cobaltite spinel $(Mn,Co)_3O_4$, (MCO) is commonly applied[5,8]. The process operations include cleaning of incoming parts followed by aerosol spray deposition of a suspension containing the MCO oxide powder, binder, dispersant and solvent followed by forced air drying. Multiple deposition-dry cycles may be necessary to build up the required coating thickness. After coating, the part is subjected to two heat treatments to develop an adherent, dense coating. The first heat treatment consist of a controlled atmosphere, reduction-firing which reduces the spinel to Co metal and MnO. During subsequent oxidation in air (or *in-situ* during stack start-up) the Co and MnO react with oxygen to reform the spinel with enhanced densification via reaction-sintering.

For the non-active, seal area coatings are necessary because chromia-forming steels interact with alkaline aluminosilicate sealant glasses to form chromates (for example, $SrCrO_4$ and $BaCrO_4$). These low thermal expansion coefficient chromates lead to failure during stack thermal cycling. Unfortunately, the active area MCO coating is not suitable. It is too chemically reactive with the sealant glasses, leading to rapid dissolution. An alternative insulating coating with low chemical reactivity is necessary for the seal perimeters. NexTech has developed two coating approaches, an oxide-based overlay protective coating and a diffusion-based aluminide coating. These coatings are compatible with the MCO coating and can be selectively deposited together by ASD and co-fired in a single heat treatment. Figure 2 shows metallic interconnects with multiple coatings applied by ASD; a MCO cathode active area coating, oxide overlay non-active area coating, and an oxide anode coating.

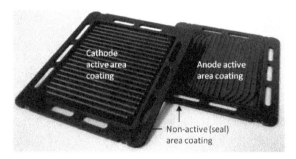

Figure 2. Example of metallic interconnects with multiple co-processed protective coatings

This paper will outline performance validation testing that has been completed on the MCO coating to evaluate oxidation and chromium volatilization resistance that limit the service lifetime of coated metallic interconnects.

EXPERIMENTAL

For evaluation of the MCO active layer coating, 10 μm thick MCO coatings were symmetrically applied by aerosol-spray deposition (ASD) to samples of AL 441HP[TM], a commercially available ferritic stainless steel. Each coating pass deposited 5 μm. Two spray/dry cycles were required to build up the required coating thickness. Coated samples were then fired, first in a controlled atmosphere reduction-firing followed by oxidation in air both at elevated temperature. The ex-situ oxidation firing was necessary to fully densify the coating. Without this treatment initial oxidation weight-gain data would be dominated by densification of the coating and not by growth of the underlying scale.

Oxidation kinetics of uncoated and MCO coated AL 441HP[TM] samples were investigated by isothermal, discontinuous weight-gain experiments in air at three temperatures: 800 °C and accelerated conditions: 900 °C and 1000 °C. Samples were removed at periodic intervals, their weights recorded, and the coating inspected. To reduce sources of error multiple samples were run for each condition and multiple weight measurements taken. The oxidation testing continued until a sample set displayed evidence of spallation.

The long-term, electrical performance of MCO coated AL 441HP[TM] samples was evaluated through electrochemical stability testing. The area specific resistance (ASR) was measured using a four-terminal dc technique. Samples were provided with LSM conductive pads for electrical contact and dried. Platinum leads and mesh were bonded to the pads with LSM ink and cured at 1000°C for one hour in air. The samples were electrically tested under an atmosphere of humidified air, at a current density of 0.5 A/cm^2 at both 800 °C and 900 °C. Post-test characterization of the coating microstructure by cross-section scanning electron microscopy and energy dispersive spectroscopy compositional analysis was used to understand how the coating microstructure evolved and allowed for an assessment of the effectiveness of the coating to act as a chromium diffusion barrier to be made.

RESULTS AND DISCUSSION

Oxidation Analysis

One of the primary functions of the protective coating is to improve oxidation resistance. Table I lists the time to spallation for uncoated and coated samples at 800, 900 and 1000 °C. For both coated and uncoated samples oxidized at 800 °C no spallation was observed. At higher temperatures spallation was observed, occurring at shorter times with increasing temperatures. At 1000 °C spallation of uncoated AL 441HP[TM] samples was extremely rapid (20 hours).

Before spallation, oxidation was found to follow parabolic kinetics, consistent with diffusion-limited oxidation as depicted in Figure 3 for a) uncoated AL 441HP[TM] and b) MCO coated AL 441HP[TM] at 800, 900, and 1000 °C. Parabolic rate constants, k_p calculated from the slope of the line of best fit are summarized in Table II. Oxidation resistance is significantly improved for coated versus uncoated samples. For all three temperatures, the rate of oxidation is reduced by approximately 10X for MCO coated versus uncoated AL 441HP[TM]. It is believed that the improvement in oxidation resistance is derived from the effectiveness of the MCO coating as a barrier to inward oxygen diffusion.

Oxidation kinetics displayed an Arrhenius temperature dependence for both coated and uncoated samples. Accelerated testing protocols for the electrical testing, based on these accelerated oxidation kinetics at higher temperatures enabled expeditious validation of the coating, as discussed in the next section.

Table I. Time to spallation for uncoated and MCO coated AL 441HP^{TM} substrates at 800, 900 and 1000 °C. *At 900 °C a small number of coated samples displayed evidence of spallation after 600-800 hours.

Oxidation temperature /°C	Coated/Uncoated	Time to spallation /hrs
800	Uncoated	-
	Coated	-
900	Uncoated	200-400
	Coated	600-800*
1000	Uncoated	20
	Coated	120

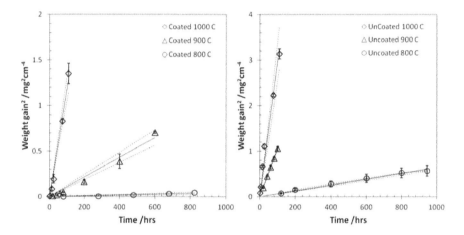

Figure 3. Weight-gain squared versus time for a) uncoated and b) MCO coated AL 441HP^{TM} substrates at 800, 900, and 1000 °C. Parabolic rate constants were calculated from the slope of the lines of best fit. The dotted lines indicate 95% confidence intervals.

Table II. Parabolic oxidation rate constants for MCO coated and uncoated AL 441HP^{TM} samples at 800, 900, and 1000°C.

Oxidation temperature /°C	Coated/Uncoated	Parabolic rate constant k_p / mg^{2}cm^{-4}s^{-1}
800	Uncoated	1.79×10^{-13}
	Coated	1.28×10^{-14}
900	Uncoated	2.90×10^{-12}
	Coated	2.82×10^{-13}
1000	Uncoated	1.58×10^{-11}
	Coated	2.95×10^{-12}

Electrical Area Specific Resistance Testing

The long-term ASR behavior of MCO coated AL 441HP™ samples is shown in Figure 4. For the first 1600 hours the samples were tested at 800 °C, consistent with standard SOFC operating temperatures. The temperature was then increased to 900 °C to accelerate thermally driven failure mechanisms before the temperature was returned to 800 °C for thermal-cycling and long-term stability testing.

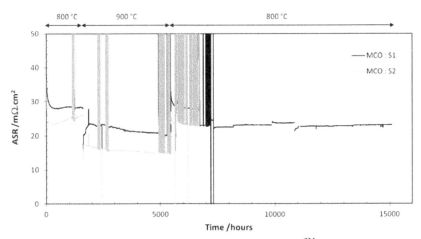

Figure 4. Long-term ASR behavior of two MCO coated AL 441HP™ substrates. Test conditions were 800 or 900 °C in humidified air with an applied current density of 0.5 A/cm². After 6700 hours sample 2 was removed from test for post-mortem microstructure characterization.

The ASR electrical testing is able to detect several failure mechanisms associated with the MCO coated interconnects directly. For example, oxidation driven growth of the resistive native chromia oxide correlates to the measured ASR. Growth of the native scale also leads to a reduction in interfacial strength and getter susceptibility of the coating to spall. Spallation leads to a reduction in the electrode active area and therefore is also readily detected in the ASR measurement. The two samples demonstrate very good performance, with low and extremely stable ASR values. The behavior indicates that the MCO coating is effective at increasing the oxidation resistance.

To investigate the integrity of the coating/scale/substrate interface both samples were subjected to multiple thermal cycles. These cycles correspond to the vertical bars in Figure 4. Thermal cycles were performed in sets of 10 from test temperature to room temperature with a ramp rate of 5 °C/min and dwell of four hours. This was subsequently changed to more aggressive conditions of 10 °C/min and dwell of one hour. Figure 5 enlarges the ASR data for the thermal cycling region at 6,000 hours, highlighting the aggressive thermal cycling. No degradation in ASR performance is observed throughout the multiple thermal cycles. The samples return to the stable ASR baseline after each set of thermal cycles.

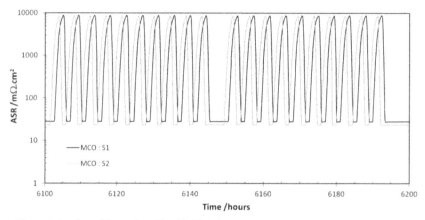

Figure 5. Section of thermal cycling highlighting sets of 10X thermal cycles from 800 °C to room temperature. The heating/cooling rates were 10 °C/min with a one hour dwell. The two samples are offset to aid viewing..

The oxidation acceleration factors determined from the calculated parabolic rate constants (Table II) can be used to estimate the effective lifetime of the electrical samples tested under accelerated conditions. The 3800 hours the samples spent at 900 °C corresponds to greater than 40,000 hours equivalent oxidation hours at 800 °C. Although care must be taken in extrapolating accelerated testing results, it suggests that the MCO coating is able to successfully suppress oxidation of metallic interconnects to allow lifetime targets to be achieved. for As the performance of the coatings continues to remain stable for ever longer times, confidence in these predictions is further reinforced.

Post-Test Characterization and Chromium Volatilization Resistance
      Chromium evaporation is a significant failure mechanism for chromium containing stainless steel interconnects as it can lead to poisoning of the fuel cell cathode and loss of performance. This failure mechanism is not able to be resolved from the ASR testing, however, careful post-test characterization of the coating microstructure enabled an assessment of the effectiveness of the coating to act as a chromium diffusion barrier.
The microstructure of a MCO coating on AL 441HP^{TM} after less than 1000 hours at 800 °C is shown in the SEM cross-section image (Figure 6.a). The corresponding chromium compositional EDS map (Figure 6.b) highlights the native chromia scale that forms at the interface between the coating and substrate. Segmentation experiments indicate that this chromia layer (thickness 2 μm-3 μm) was primarily established during the oxidation heat treatment. Compositional scans within the MCO coating indicate that chromium is restricted to this chromia layer with very little chromium diffusion into the MCO coating. Weight percent chromium midway through the coating is less than 1.5 % and no chromium is detected in the outer half of the coating.
To understand how the coating microstructure evolves with ageing, after 6700 hours on long-term ASR testing sample 2 (Figure 4) was removed and its microstructure characterized by cross-section SEM (Figure 7.a) and compositional EDS analysis (Figure 7.b). The coating microstructure is relatively unchanged from the microstructure in Figure 6, indicating that the coating is stable at these operating temperatures (800 and 900 °C). Chromium is again primarily

constrained within the chromia scale at the coating/substrate interface. There is only limited diffusion of chromium within the coating. The thickness of the chromia scale has increased but is still only 4 μm-5 μm. There has therefore been only limited growth of the resistive native chromia scale demonstrating that the MCO coating is successful at improving oxidation resistance of the substrate. The level of chromium within the coating (center of coating: chromium weight percent is 13 % reducing to 3 % in the outer region of the coating) is higher than in Figure 6 which indicates that there has been some outward diffusion of chromium into the coating.

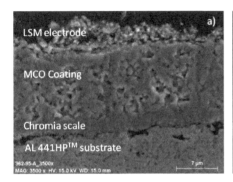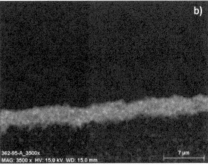

Figure 6. a) Cross-section SEM image and b) chromium composition EDS map of a MCO coating on AL 441HP^{TM} after 800 hours of testing at 800 °C in humidified air with an applied current density of 0.5 Acm^{-2}.

Interestingly, in addition to outward diffusion of chromium from the native scale into the coating there does appear to be some diffusion of Mn (and Co) from the coating inward, into the scale. This leads to the possibility of the MCO coating doping the native chromia scale and leading to an improvement in the conductivity of the scale which in turn could improve ASR performance of the coated interconnect.

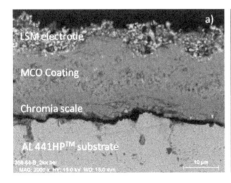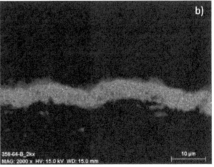

Figure 7. a) Cross-section SEM image and b) chromium composition EDS map of a MCO coating on AL 441HP^{TM} after 7600 hours of testing at 800/900 °C in humidified air with an applied current density of 0.5 Acm^{-2}.

CONCLUSION
The effectiveness of a low-cost aerosol-spray deposition coating process for applying protective oxide coatings to metallic interconnects for SOFC applications has been demonstrated. Oxidation resistance, electrical stability and chromium volatilization of MCO coated AL 441HP^{TM} ferritic stainless steel have been evaluated under SOFC operating conditions. Results indicate that the MCO coating is successful at improving oxidation resistance and chromium volatilization resistance and that the selective application of protective coatings to metallic interconnects is an encouraging strategy to satisfy the required lifetime performance requirements.

ACKNOWLEDGEMENT
Financial support from the Department of Energy SBIR program (DE-PS02-08ER08-34) is gratefully acknowledged

REFERENCES
1 N. Shaigan, W. Qu, D. G. Ivey, and W. Chen, A Review of Recent Progress in Coatings, Surface Modifications and Alloy Development for Solid Oxide Fuel Cell Ferritic Stainless Steel Interconnects, *J. Power Sources*, **195** 1529-42 (2010).
2 J. W. Fergus, Metallic interconnects for solid oxide fuel cells, *Mat. Sci. Eng. A*, **397** 271-283 (2005)
3 W. J. Quadakkers, V. Shemet, and L. Singheiser, High Temperature Material, US Patent No. 2003059335 (2003)
4 ThyssenKrupp VDM GmbH, Crofer 22APU data sheet No. 4146 (2006)
5 T. Uehara, N. Yasuda, M. Okamoto, and Y. Baba, Effect of Mn-Co Spinel Coating for Fe-Cr Ferritic alloys ZMG232L, and 232J3 for Solid Oxide Fuel Cell Interconnects on Oxidation Behavior and Cr-Evaporation, *J. Power Sources*, **196**, 7251-56 (2011)
6 Z. Yang, G. Xia, C. Wang, Z. Nie, J. Templeton, J. Stevenson, and P. Singh, Investigation of AISI 441 Ferritic Stainless Steel and Development of Spinel Coatings for SOFC Interconnect Applications, PNNL Report 17568, (2008).
7. Z. Yang, G. –G. Xia, G. D. Maupin, and J. W. Stevenson, Conductive Protection Layers on Oxidation Resistant Alloys for SOFC Interconnect Applications, Surf. Coat.Tech., 201 4476-83 (2006)
8 Z. Yang, G. –G. Xia, X. –H. Li, and J. W. Stevenson, (Mn,Co)3O4 spinel coatings on ferritic stainless steels for SOFC interconnect applications, Int. J. Hydrogen Energ., 32, 3648-54 (2007)

VISCOUS SEALING GLASS DEVELOPMENT FOR SOLID OXIDE FUEL CELLS

Cheol Woon Kim[1], Jen Hsien Hsu[2], Casey Townsend[2], Joe Szabo[1], Ray Crouch[1], Rob Baird[1], and Richard K. Brow[2]

[1]MO-SCI Corporation, Rolla, MO, USA
[2]Department of Materials Science and Engineering and the Graduate Center for Materials Research, Missouri University of Science and Technology, Rolla, MO, USA

ABSTRACT

Glass compositions have been formulated and tested for use as viscous seals for solid oxide fuel cells (SOFCs). These alkali-free borosilicate glasses possess desirable thermo-mechanical properties and thermo-chemical characteristics, and exhibit promising hermetic sealing and self-healing behavior under SOFC operational conditions. The dilatometric softening points (T_s) and the glass transition temperatures (T_g) of the glasses are generally under 650°C, the lower bound of the SOFC operating temperature. To date, glass seals between a YSZ-NiO/YSZ bilayer and aluminized stainless steel 441 have survived 100 thermal cycles (750°C to room temperature) in dry air at a differential pressure of 0.5 psi (26 torr) over the course of > 3,300 hours without failure, and 103 thermal cycles under wet forming gas. Seals intentionally cracked upon quenching from 800°C to RT at >25°C/s become hermetic upon reheating to 700°C and higher.

INTRODUCTION

Solid oxide fuel cells (SOFCs) require hermetic seals to prevent mixing of the fuel and oxidant streams within the cell stack and to seal the stack to the system manifold. Reliable sealing materials must (a) possess viscosity-temperature characteristics that are compatible with sealing requirements and that allow for stress relaxation and 'self-healing' without excessive flow, under pressure, that would compromise seal integrity; (b) be chemically compatible with SOFC components and so not alter the long-term thermo-mechanical stability of the seal by forming deleterious interfacial reaction products; (c) avoid the significant volatilization of glass constituents under the SOFC operational conditions that has been associated with other sealing materials, and so alter the viscous properties of the seal or the performance of the SOFC; (d) possess dilatometric properties that are compatible with other SOFC components, and these properties must be stable over the course of the SOFC lifetime ($\geq$ 40,000 hour).

The use of 'viscous glass' seals provides one means of reducing the risk that thermal stresses will result in catastrophic failures, and may provide a means for the seal to 'recover' if cracks do form. Some of the early work at Argonne National Laboratory identified compositions in the alkaline earth lanthanum aluminoborate system that remained viscous ($\eta > 10^3$ Pa-s) under operational conditions [1,2]. More recently, Singh [3,4] has proposed that SOFC seals be made from stable (non-crystallizing) glass compositions with glass transition temperatures (T_g) well-below the SOFC operational temperature. On heating the seal above T_g, the rigid glass becomes viscous and any flaws within the seal (or at a seal interface) will 'heal' because of viscous flow. In addition to providing a means to 'repair' cracks in a seal caused by thermal stresses, the use of a viscous seal could also reduce the magnitude of those stresses compared with a rigid glass seal with the same thermal expansion characteristics since the stresses will be relieved at temperatures above T_g in the 'viscous glass' seal, reducing the effective ΔT over which thermal stresses develop.

RESULTS AND DISCUSSION

Glass Formulation and Properties

New glass compositions were formulated and these alkali-free glasses have compositions from the $BaO-B_2O_3-SiO_2-Al_2O_3-RO$ system where RO represents other alkaline earth oxides or ZnO. The glasses were melted in fused silica crucibles (99.6% SiO_2, 0.2% Al_2O_3) in ambient air for 2-7 hours, typically at 1050-1100°C. The dilatometric softening points (T_s) and the glass transition temperatures (T_g) of the glasses are generally close to 650°C, the lower bound of the SOFC operating temperature. The glasses generally do not crystallize in a differential scanning calorimeter (DSC) when heated at a rate of 10°C/min up to 1000°C (Figure 1). The coefficient of thermal expansion (CTE) values (40-500°C) of Glass 73, 75, and 77 are $8.5 \times 10^{-6}/°C$, $8.2 \times 10^{-6}/°C$, and $9.3 \times 10^{-6}/°C$ respectively. The liquidus temperatures (T_L) of Glass 73, 75, and 77 are 800±10°C, 810±10°C, and 810±10°C respectively (Figure 2) and so these glasses can form viscous seals that do not substantially devitrify under SOFC operational conditions. The T_L measurements were conducted per ASTM C 829-81 [5].

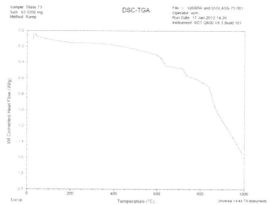

Figure 1. DSC analysis of the selected viscous sealing glass (Glass 73); there are no obvious crystallization exotherms up to 1000°C.

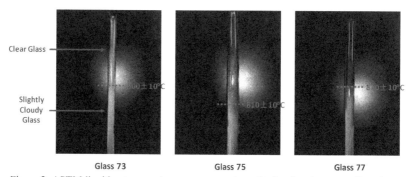

Figure 2. ASTM liquidus temperature measurement results for the viscous sealing glasses showing no obvious crystallized phases after 72 hours at the SOFC operational temperature.

Viscosity

The viscosities of candidate glass melts were measured at intermediate temperatures using a cylinder compression technique with a dynamic-mechanical analyzer (Perkin-Elmer DMA-7) and at high temperatures using a rotating spindle technique with a Haake High Temperature Viscometer (ME 1700). For the cylinder compression method, data were collected using two applied loads, 500 mN and 2000 mN, extending the range of viscosity measurements to lower temperatures. Figure 3 summarizes the viscosity data collected for Glass 73, 75, and 77, with their respective fits to the Corning Viscosity Model (CVM) [6]. This new CVM is based on fitting parameters that have physical meaning and is reported to provide more accurate predictions for extrapolated η-T curves than do other models:

$$\log\eta(T) = \log\eta_\infty + (12 - \log\eta_\infty)\frac{T_g}{T}\exp\left[\left(\frac{m}{12-\log\eta_\infty} - 1\right)\left(\frac{T_g}{T} - 1\right)\right] \qquad (1)$$

Here, T is temperature in K, η is the viscosity in Pa-s, η_x is the high-temperature limit of viscosity, T_g is the glass transition temperature defined as the temperature at which the glass viscosity is 10^{12} Pa-s, and m is the fragility parameter, a measure of the non-Arrhenius nature of the temperature dependence of viscosity. For the present fits to the CVM equation, the value of $\log\eta_x$ was held constant at -3.5 and the other fitting parameters (m and T_g) were optimized based on the experimental results. Singh [4] reports 'self-healing' behavior for an SOFC sealing glass with a viscosity of 10^5 Pa-s at 800°C. From Figure 3, similar behavior would be expected for Glass 73, 75, and 77 at temperatures < 800°C. The Littleton softening point ($10^{6.6}$ Pa-s) is sometimes defined as the temperature at which a glass will flow under its own weight. With this definition, self-healing behavior should be possible at temperatures above about 675°C.

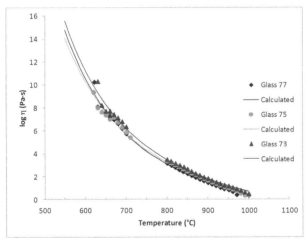

Figure 3. Viscosity data for candidate glasses and the corresponding CVM fitted lines.

Holding Glass 73 at 800°C for up to 1,000 hours had little effect on the viscosity characteristics of the melt, in the viscosity range of 10^6 to 10^{10} Pa·s (Figure 4), despite the presence of some crystals in these samples (Figure 5). After 2,000 hours, however, there is a change in the viscosity-temperature characteristics. For this sample (the black circles in Figure 4), less viscous flow occurs up to 900°C. This behavior can be explained by the crystallization in this sample (Figure 5). It appears that when the 5 mm cylinder is compressed during the viscosity measurement, interlocking crystals in the sample inhibit the downward progress of the probe compared to the normal viscous flow of a softening glass. Similar constrained viscosity-temperature behavior is noted in Figure 4 for the crystallized sample held in air (dry) at 750°C for 2,000 hours.

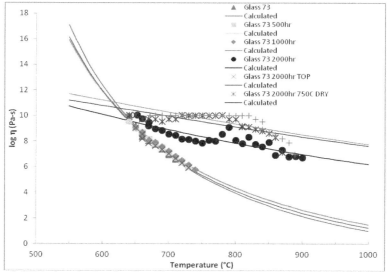

Figure 4. DMA viscosity data for Glass 73 after different thermal histories; the solid lines are fits based on the Corning Viscosity Model.

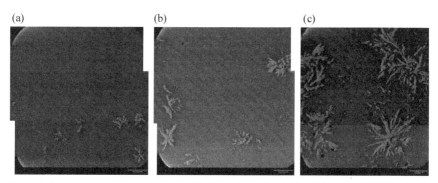

Figure 5. Scanning electron microscopy (SEM) micrographs of Glass 73 held at 800°C in air for (a) 500, (b) 1,000, and (c) 2,000 hours.

Hermeticity and Self-Healing Behavior

Coupon sealing was conducted by sandwiching Glass 73 paste between an aluminized SS441 disc (3.2 cm diameter and 1 mm thick) with a central hole (1 cm diameter) and an anode-supported (NiO/YSZ) thin electrolyte (YSZ) bilayer square (2 cm side) as shown in Figure 6. The SOFC materials, aluminized SS441 and YSZ-NiO/YSZ bilayers, were supplied by Pacific Northwest National Laboratory (PNNL). The seal sandwich assembly was heated at 10°C/min to 850°C and held for 8 hours to remove binder and bubbles and to form the desired seal. The assembly was then transferred to a pressurized tube furnace and exposed to thermal cycles between 750°C and room temperature (RT) in either dry air or wet forming gas (5% H_2 and 95% N_2 plus water) at a differential pressure of 0.5 psi (26 torr). The Glass 73 seal has survived 100 thermal cycles (750°C-RT) in dry air over the course of > 3,300 hours without failure. The Glass 73 seal has also survived 103 thermal cycles under wet forming gas. The results are summarized in Figures 7 and 8, and both tests were deliberately terminated for analysis.

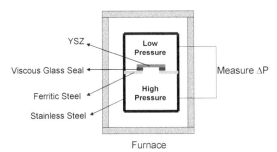

Figure 6. Schematic of the seal performance test system.

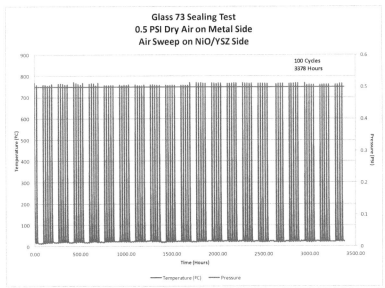

Figure 7. Demonstration of a hermetic seal made from SS441/Glass 73/NiO-YSZ bilayer. This seal has survived 100 thermal cycles (750°C-RT) in dry air at a differential pressure of 0.5 psi (26 torr) over the course of more than 3,300 hours. The test was deliberately terminated for analysis.

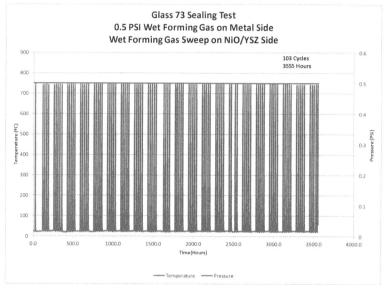

Figure 8. Demonstration of a hermetic seal made from SS441/Glass 73/NiO-YSZ bilayer. The seal has survived 103 thermal cycles (750°C-RT) under wet forming gas at a differential pressure of 0.5 psi (26 torr) over the course of more than 3,500 hours. The test was deliberately terminated for analysis.

SOFC Reaction Couples

Figure 9 displays representative SEM images at different magnifications of the sandwich sample after the long-term thermal cycle test (100 cycles in dry air). The glass cracked along the metal interface while the SEM sample was being prepared, and the epoxy that was used to mount the samples filled the interface between the glass layer and the SS441. Crystals and a few voids have formed in the glass layer. No Zr-rich crystals were observed at the glass/YSZ interface.

Figure 9. Micrograph of a Glass 73 sandwich seal that remained hermetic after 100 thermal cycles (750°C to RT) in dry air. The crack in the image on the left was formed during the preparation of the SEM sample.

Although the crack is close to the glass/metal interface, Al- and Si-rich needle-shape crystals were observed (Figure 10(a)). The source of the aluminum in these crystals may be the Al-rich scale on the SS441 surface. The reasons for this hypothesis include the observation that no Al-rich crystals are observed in Glass 73 heated alone at 800°C, and the observation that after short times, the Al-rich phases are concentrated near the metal/glass interface. The influence of those crystals on the long-term sealing function of the glass is still unknown.

Crystals (areal fraction: ~17%) formed in the glass layer of the thermally cycled sample (Figure 9). Two types of crystals were detected. One is Si-rich and plate-like, similar to those that form when Glass 73 is crystallized. The second type is the Al- and Si-rich needle-shaped crystal which might be related to the interaction between glass and the Al-rich coating on the SS441.

There is little evidence for interactions between Glass 73 and YSZ (Figure 10(b)). No Zr- or Y- containing crystals were detected at the glass/YSZ interface. The crystals which have accumulated near the interface are similar to ones detected elsewhere in glass layer.

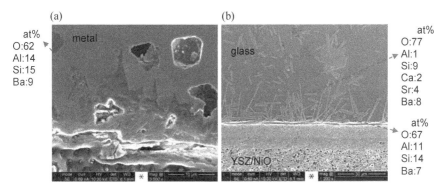

Figure 10. Micrograph of (a) the SS441/glass interface; and (b) the glass/YSZ interface of a Glass 73 sandwich seal that survived 100 thermal cycles (750°C to RT).

Volatility

Glass stability against volatilization was determined by weight loss measurements at elevated temperatures. Weight loss measurements were conducted as a function of time (up to 2,000 hours) at 750°C and 650°C in flowing wet reducing conditions (5% H_2 and 95% N_2 with a flow rate of 10 mL/s) and stagnant dry air conditions. The forming gas was bubbled through deionized water held at 70°C so that the atmosphere contained ~30 vol% water. Glass samples were periodically removed from the furnace (Figure 11), cooled in a desiccator, and measured for weight change per unit area. Figure 12 shows the linear volatility of Glass 73 in different atmospheres at 750°C and 650°C for up to 2,000 hours. The volatilization rates are 2.0×10^{-8} g/mm^2/hr under flowing wet reducing conditions and 1.7×10^{-8} g/mm^2/hr under stagnant dry air conditions, respectively at 750°C; and 1.4×10^{-8} g/mm^2/hr under flowing wet reducing conditions at 650°C. The inductively coupled plasma mass spectroscopy (ICP-MS) analysis on the water trap downstream from the heat-treated glass samples indicates that boron is the only component volatilized from the glasses.

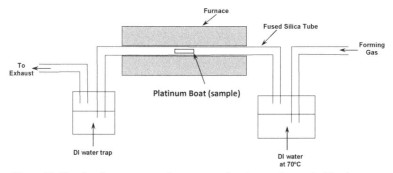

Figure 11. Simulated steam atmosphere system for viscous glass volatilization test.

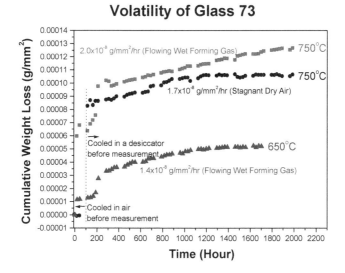

Figure 12. Volatility measurements for Glass 73 at 750°C and 650°C in different atmospheres.

Crack Healing Experiments

Crack healing through viscous flow is commonly observed in glass at elevated temperatures and is one important reason why a viscous glass was chosen as the sealing material. Self-healing of glass cracks caused by thermal shock was demonstrated using Glass 73 in a sandwich-sealed coupon, schematically shown in Figure 13(a). The hermeticity of sealed samples is checked using the pressure test assembly (Figure 13(b)). The hermetic sample is then re-heated to 800°C, then rapidly quenched (>25°C/s) to thermally shock the glass, producing visible cracks through the glass layer (Figure 14(a)). When this cracked sample was re-heated to 800°C for 2 hours and slowly cooled (~16°C/min) back to room temperature, the sample was again hermetic, holding a 2 psi differential pressure, and the visible crack was healed (Figure 14(b)). This is the first demonstration of 'crack healing' in a thermally shocked seal.

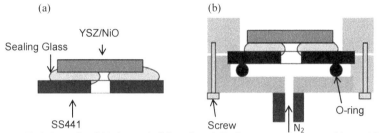

Figure 13. Schematic of (a) the sandwich seal, and (b) the pressure test assembly used in the crack healing experiments.

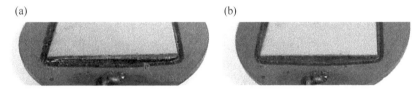

Figure 14. Sandwich seal coupon made with Glass 73; (a) crack caused by high cooling rate quench, (b) crack healed after re-heating to 800°C for 2 hrs.

Experiments were repeated to verify the self-healing of Glass 73 sealing at 800°C, and additional experiments were performed to determine if self-healing would occur at lower temperatures. A hermetic seal is initially heated to 800°C, removed from the furnace and then the glass is cracked by quenching one edge of the metal disk with a stream of water. The cracked seal was re-heated to the designated temperature for two hours and slowly cooled to room temperature where hermeticity was then evaluated. If the cracked glass re-sealed and passed the hermeticity test, the sample was tested again, but at a lower "re-sealing" temperature. So far, self-healing of cracked seals has been observed multiple times when samples were re-heated to 725°C for 2 hours (Figure 15). When the healing temperature was 700°C, a cracked seal was re-healed once, but not a second time (Table 1). From these experiments, the lowest self-healing temperature for Glass 73 sealing might be close to 700°C. From the viscosity-temperature curve (Figures 3 and 4), the viscosity of Glass 73 at 700°C is $10^{6.8}$ Pa-s, close to the Littleton softening point ($10^{6.6}$ Pa-s), sometimes described as the temperature at which a glass flows under its own weight.

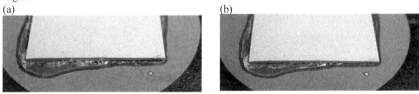

Figure 15. Sandwich coupon made with Glass 73; (a) crack caused by high cooling rate quench, (b) crack healed after re-heating to 725°C for 2 hrs.

Table 1. Summary of self-healing experiments with Glass 73.

Temperature	Time (hr)	Viscosity, log η (Pa-s)	Observation (# of experiments)
800°C	2	3.7	Healed (6 tests)
750°C	2	5.1	Healed (2 tests)
725°C	2	5.9	Healed (3 tests)
700°C	2	6.9	Healed once, but not a second time

CONCLUSION

We identified and tested several glass compositions that could be used as viscous seals for SOFCs. These alkali-free borosilicate glasses possess desirable viscosity characteristics, relatively low liquidus temperatures and minimum reactivity with the SOFC components, exhibit no significant weight losses from the molten state when held in dry air and flowing wet reducing condition, and exhibit promising hermetic sealing and self-healing behavior under SOFC operational conditions. To date, glass seals between a YSZ-NiO/YSZ bilayer and aluminized SS441 have survived 100 thermal cycles from 750°C to room temperature in dry air at a differential pressure of 0.5 psi over the course of > 3,300 hours without failure, and 103 thermal cycles under wet forming gas. We have successfully demonstrated crack healing in thermally cracked glass seals, for the first time, at temperatures as low as 700°C.

ACKNOWLEDGEMENT

This work was supported by the Small Business Innovation Research (SBIR) program of the US Department of Energy (DOE) under contract DE-SC0002491. The authors would like to thank Dr. Joseph Stoffa for his assistance and Drs. Yeong-Shyung Matt Chou and Jeff Stevenson for their SOFC material sample supply and discussion.

REFERENCES

[1] I. Bloom and K. L. Ley, "Compliant Sealants for Solid Oxide Fuel Cells and Other Ceramics," US Patent 5,453,331, issued Sep. 26, 1995.
[2] K. L. Ley, M. Krumpelt, R. Kumar, J. H. Meiser, and I. Bloom, "Glass-ceramic sealants for solid oxide fuel cells: Part I. Physical properties," *J. Mat. Res.*, 11 (1996), 1489- 1493.
[3] R. Singh, "High temperature seals for solid oxide fuel cells," *Proceedings of SSM International Conference*, Columbus, OH, Oct. 18-20, 2004
[4] R. Singh, "Innovative seals for solid oxide fuel cells (SOFC)," Final Progress Report (2008), DOE Award DE-FC26-04NT42227.
[5] ASTM–American Society for Testing and Materials, "Standard Practices for Measurement of Liquidus Temperature of Glass by the Gradient Furnace Method," ASTM C 829-81 (Reapproved 2010), ASTM International, West Conshohocken, Pennsylvania.
[6] J. C. Mauro, et al. (2009), *PNAS*, 106 (47), 19780-4.

PROPANE DRIVEN HOT GAS EJECTOR FOR ANODE OFF GAS RECYCLING IN A
SOFC-SYSTEM

Andreas Lindermeir, Ralph-Uwe Dietrich, and Christoph Immisch
Clausthaler Umwelttechnik Institut GmbH (CUTEC)
Clausthal-Zellerfeld, Germany

ABSTRACT
A compact propane driven SOFC-system with recycling of hot AOG is developed at
CUTEC Institute with partners from Zentrum für BrennstoffzellenTechnik GmbH (ZBT,
Duisburg), Institute of Energy and Process Systems Engineering (InES, TU Braunschweig)
and Institute of Electrical Power Engineering (IEE, TU Clausthal). The system extends the
commercially available integrated stack module (ISM) of Staxera GmbH (Dresden,
Germany) by the required fuel processing unit and auxiliary components. The expected
electrical power output for a propane flow of 1.0 l_N/min is 950 W_e (gross). Thus, electrical
system efficiency will be 61 % (based on propane LHV).

CUTEC developed a tailor-made hot gas ejector for anode offgas recycling that uses
pressurised propane from standard gas bottles as propellant gas. Propane leaves the ejector
nozzle at high velocity and hereby mixes with AOG. A Laval nozzle accelerates the propane
stream to supersonic speed and enables a recycle ratio sufficient for soot-free reformer
operation. As the ejector has no moving parts it is expected to work robust, even at the high
operating temperatures of about 600 °C.

The system concept, design options for thermal integration and high compactness and
experimental results for the component development will be discussed. Ejector performance
data will be presented.

INTRODUCTION
SOFC-systems with either internal or external reforming allow power generation from
common hydrocarbon fuels like natural gas, LPG or diesel. Especially propane is easy to
handle and widely used in camping and leisure applications. Because commercially available
SOFC stacks are not yet suited for direct oxidation of hydrocarbon fuels, different approaches
for the external reforming are considered today, e.g. steam reforming (SR) using water or
partial oxidation (POX) using air-oxygen. However, these concepts suffer either from
complex auxiliary units for the water conditioning or low electrical system efficiency.

A highly efficient alternative is the reforming of hydrocarbon fuels with the anode off
gas (AOG) of the SOFC, promising electrical system efficiencies above 60 %. Recycling a
part of the AOG supplies the reformer with the SOFC oxidation products steam and carbon
dioxide as oxygen carriers. Fuel conversion to hydrogen and carbon monoxide via combined
steam- and dry-reforming yields a higher chemical energy fed to the stack compared to the
POX reforming. Heat required for the endothermic steam- and dry-reforming of propane fuel
can be provided by combustion of the remaining AOG in the burner and heat transfer to the
reforming reactor.

In a previous project it was demonstrated that AOG-recycling for reforming enables
higher efficiencies than POX for a propane driven SOC system operated in a controlled
furnace environment[1,2,3]. However, an autarkic operation without external heat supply is
essential for a future product development and requires a high degree of thermal integration
of all components. Development tasks to achieve an thermal self-sustaining SOFC-system
with AOG recycling are the characterization of a new generation of SOFC stacks (provided
by Staxera) and the improvement of the ejector design to achieve sufficient recycle ratios. A
Laval nozzle was designed to improve the ejector performance. The ejector has to work at
temperatures of about 600 °C to reduce energy losses.

SYSTEM CONCEPT

    The system concept has only two input streams for air and propane and one common output stream. Air is used as oxygen carrier for the cathode and the burner and serves in addition as cooling medium for the stack. As propellant gas for the ejector the already pressurised propane fuel is used. To reduce heat up time during system start up an additional propane supply to the burner is integrated, enabling supplementary energy supply. Nevertheless, at steady state operation propane is supplied only to the ejector. The only output stream is the exhaust gas from the burner. Figure 1 shows the flowchart.

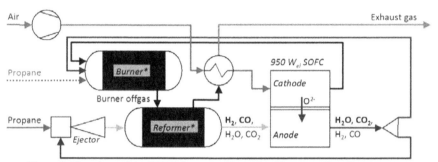

Figure 1. Flowchart of the SOFC-System with anode offgas recycling (*components developed by project partner ZBT)

    By acceleration of propane in the primary nozzle a suction pressure is generated what intakes AOG into the ejector. The mixture stream enters the reformer and propane is reformed with the steam and carbon dioxide contained in the AOG. The reformate gas enters the SOFC stack and is electrochemically oxidized at the anode. Non-recycled AOG is combusted in the catalytic burner for supplying heat to the endothermic reforming and pre-heating the cathode air in a heat exchanger by the exhaust gas. Ejector, reformer and burner are tightly arranged in one thermally insulated housing to reduce heat losses gain a small pressure drop. The housing is directly attached to the ISM casing. Bellow compensators in the tube connections between the units were used to avoid thermal stress during heat-up or shut-down. The heat exchanger is insulated separately to avoid thermal interaction with the hot components. Figure 2 shows a 3D-sketch of the system.

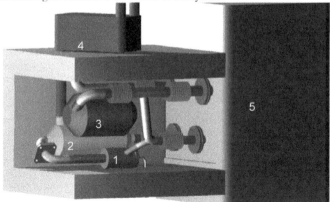

Figure 2: 3D sketch of the SOFC-system: ejector (1), reformer (2), burner (3), heat exchanger (4), integrated stack module (5)

STACK CHARACTERIZATION

The Staxera ISM contains two Mk200 stacks with 30 ESC4*-cells each. The stack was characterized in a controlled furnace environment prior to the system development to understand its performance at AOG reformate operation. ZBT conducted thermodynamic simulations of reformate composition for different fuel compositions depending on recycle-ratio, fuel utilizations of the SOFC stack, AOG recycle rates and reformer temperatures. Corresponding gas mixtures were fed to an Mk200/ESC4 stack and power output and fuel utilization as characteristic performance values were experimentally determined. The test results for 19 different fuel gas compositions and flow rates are shown in Figure 3.

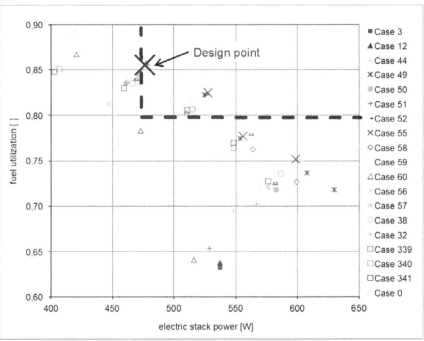

Figure 3: Characteristics of one Mk200/ESC4 stack, fuel utilization as function of stack power, designated operating range is marked by red doted lines

The designated electrical stack power of ≥ 475 W at a fuel utilization ≥ 0.8 is achieved for several operation points. Table 1 shows the calculated data for the proposed system design-point for an ISM containing two Mk200/ESC4 stacks.

Table 1: System design point (calculated data)

Propane feed	1.0 l_N/min
Recycle ratio	7
Fuel utilization	85 %
Electrical stack output	950 W
System efficiency gross	61 %

EJECTOR DEVELOPMENT

Nozzle design

A Laval nozzle was chosen to achieve high impulse energy of the propellant gas. The characteristic convergent-divergent shape of a Laval nozzle enables acceleration of a gas stream to supersonic speed (see Figure 4.)

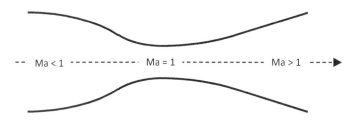

Figure 4: Convergent-divergent shape of a Laval nozzle

The correlation between the cross-section and the x-coordinate can be expressed as

$$\frac{dA}{dx} = \frac{1}{c}\frac{dc}{dx}\frac{Ma^2 - 1}{1}A \tag{1}$$

with A as cross-section area, x as coordinate along the center line, the Mach number Ma and c for the gas velocity. Assuming acceleration of gas stream ($dc/dx > 1$) the term dA/dx is negative for $Ma < 1$ thus a converging cross section results. dA/dx is positive for $Ma > 1$, resulting in a diverging cross-section area[5].

Nozzle prototypes with a Laval-shape design were produced by die sinking for the experimental ejector evaluation.. Figure 5 shows a cross-section of the negative cast of one nozzle. The maximum propane flow rate is limited by the smallest nozzle diameter and was determined to 1 l_N/min for an inlet pressure of 3.5 bar(a) and an outlet pressure of 1.02 bar(a).

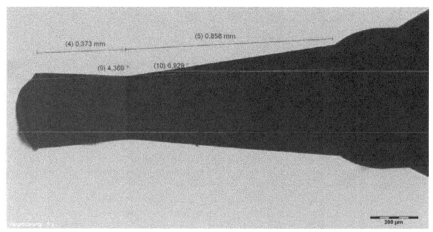

Figure 5: Cross-section of negative cast of a Laval nozzle, gas flow from right to left

Ejector design

The suction chamber of the ejector was designed without sharp edges to avoid energy dissipation and thus increase the achievable recycle ratio. The suction pipe is connected to the ejector at an flat angle to reduce inlet losses caused by the flow redirection. The diffusor section is formed with a flat angle of 5° to avoid stalls[6]. Spacer rings between the nozzle and the nozzle mount were fabricated to determine the optimum distance between the nozzle outlet and the mixing chamber inlet. Figure 6 shows a 3D-cross section of the ejector with the nozzle mounted at the reference position ±0 mm. Positions of -2.5 mm and -5.0 mm are adjusted by the spacers and investigated as well. An offset of +1.5 mm is realized by cutting the nozzle shoulder. To investigate the influence of the mixing channel diameter, different cartridges can be used to constrict the diameter of the mixing channel from 4.5 to 3.0 mm and 2.0 mm. All ejector parts were made of high temperature steel (1.4828). The Institute of Mechanical Engineering (IMW) of the TU Clausthal did the die sinking of mixing channel and diffusor.

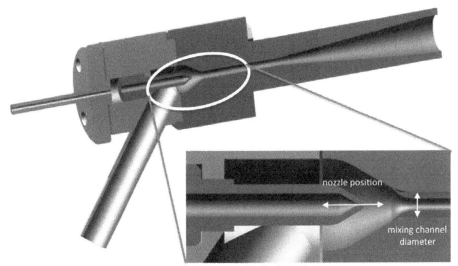

Figure 6: 3D-cross section and detail of the ejector

EXPERIMENTS AND RESULTS

Cold ejector tests

The ejector was characterized at a test bench that was equipped with a needle valve in the ejector off gas pipe to simulate the pressure drop of reformer and stack and enable a controlled back-pressure. Nitrogen was used as suction gas because of its comparable density with the expected AOG. Resulting recycle ratios as function of the adjusted back-pressure are shown in Figure 8. To investigate the influence of the ejector geometry mixing channel diameters between 2 and 4.5 mm and different nozzle positions were tested.

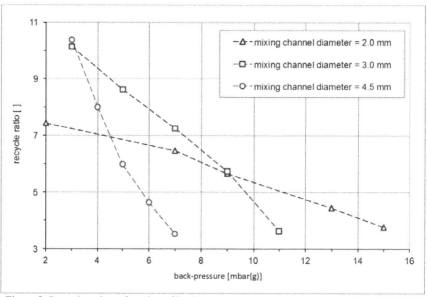

Figure 8: Recycle ratio as function of back-pressure for different mixing channel diameters, nozzle position: ± 0.0 mm; propellant gas flow rate: 1.0 l_N/min C_3H_8

The recycle ratio decreases with increasing back-pressure probably due to backflows as result of an oversized mixing channel. The highest back-pressure is achieved with the 2.0 mm diameter mixing channel but recycle ratios at lower pressure are smaller compared to larger diameters. The expected system pressure drop to be exceeded by the ejector is at least 10 mbar(g). For that back-pressure the 2.0 mm mixing channel provides best results and the smallest slope, what is beneficial for ejector operation in the system context because changes in the pressure drop have less influence on the system stability.

Additional tests were done to determine the influence of the gap between the nozzle outlet and the mixing channel inlet (see Figure 9).

Results indicate that the nozzle mounted in the reference position (±0.0 mm) shows the best performance. For larger distances between nozzle outlet and mixing channel inlet the cone shaped jet strikes the wall of the mixing chamber and momentum loss results[7,8]. For smaller distances the nozzle reduces the free cross section area and restricts the suction flow.

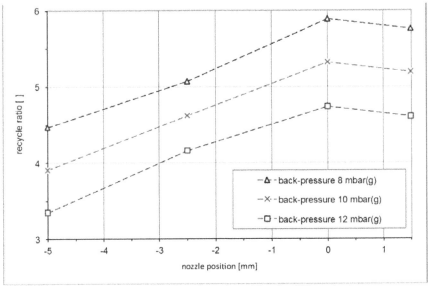

Figure 9: Recycle ratio as function of nozzle position for different back-pressure; mixing channel diameters = 2.0 mm; propellant gas = 1.0 l_N/min C_3H_8

Hot ejector tests

The influence of temperature on the ejector performance was tested in a furnace. Figure 10 shows the results for nozzle inlet temperatures between 23 °C and 693 °C.

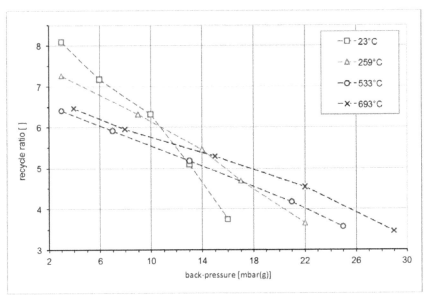

Figure 10: Recycle ratio as function of back-pressure for different nozzle inlet temperatures; mixing channel diameter: 2.0 mm; nozzle position: ± 0.0 mm; propellant gas flow rate: 1.0 l_N/min C_3H_8

For elevated temperatures, the recycle ratio is less dependent on the back-pressure and higher values are achieved for the back-pressure. At 20 mbar(g) the highest recycle ratio was 4.7. However, the demanded recycle ratio of 7 at 10 mbar(g) back-pressure was not reached, so the ejector needs further improvements.

Thermal decomposition of propane in the fuel supply has to be avoided for a stable ejector operation because solid depositions will immediately plug the nozzle and thus cut fuel supply. Tests with propane fed through the ejector nozzle at different ejector temperatures are conducted to determine the stability limit for the propane fuel. Propane decomposition can be detected by measuring the concentration of the decomposition products hydrogen and methane in the exhaust gas. Results of these measurements are shown in Figure 11, the concentrations were measured by gas chromatography.

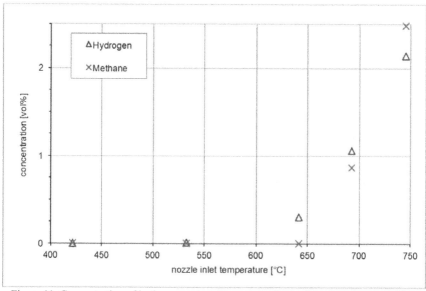

Figure 11: Concentration of hydrogen and methane as products of propane decomposition for different nozzle inlet temperatures

Neither hydrogen nor methane was detected for temperatures up to 533 °C. At 642 °C hydrogen but no methane is detected, indicating initial propene (C_3H_6) formation. At 693 °C both hydrogen and methane are found in the ejector off gas. Nevertheless, at this temperature the ejector operated continuously for about 50 minutes without increase of primary pressure or decrease of the propane flow rate. At 745 °C the nozzle inlet pressure rises and the propane flow collapses during 5 minutes because of soot formation and plugging of the nozzle. As result, ejector operation with propane is limited to temperatures below 650 °C.

SUMMARY AND OUTLOOK

An SOFC-system with anode off gas recycling is presented. The stack characterization with fuel gas derived from propane reforming with anode off gas shows that 950 W electrical power output at a fuel utilization above 0.8 is feasible with an Staxera ISM module. In conjunction with the reformer efficiency this promises a system gross efficiency of 61 %. For the recycling of the anode offgas an ejector with propane as propellant gas was designed. Laval-shaped nozzles are fabricated and the influence of geometrical parameters like mixing channel diameter and nozzle position are evaluated to improve ejector performance. Nozzle position and mixing channel diameter are set according to the required recycle ratio and back-pressure of the SOFC-system. The temperature influence on ejector performance is small compared to the geometrical factors. Propane decomposition has to be considered for temperatures above 650 °C.

Build up of the system is in progress; first results for the system operation are expected for the beginning of 2013.

ACKNOWLEDGEMENT

The authors would like to thank the colleagues from the project partners ZBT for the stationary flow sheet simulations and the development of the reformer and afterburner unit, InES for their work on the dynamic flow sheet simulations and IEE for their contribution concerning the process control and automation.

This work was financed with funds of the German "Federal Ministry of Economics and Technology" (BMWi) by the "Federation of Industrial Research Associations" (AiF) within the program of "Industrial collective research" (IGF-project no. 16638 N). Funding received from "DECHEMA Gesellschaft für Chemische Technik und Biotechnologie e.V." (Society for Chemical Engineering and Biotechnology) is gratefully acknowledged.

*ESC4 is a product name of electrolyte supported SOFC cells from H.C. Starck Ceramics GmbH[4]

REFERENCES
[1]Dietrich, R.-U., Oelze, J., Lindermeir, A., Carlowitz, O., Spitta, C., Steffen, M., Schönbrod, B., Heinzel, A., Stagge, H., Beck, H.-P., Schlitzberger, C., Chen, S., Mönnigmann, M., Leithner, R., SOFC-Brennstoffzelle mit partieller Anodenabgas-Rückführung zur Reformierung, VDI Wissensforum, 6. Fachtagung Brennstoffzelle, Braunschweig, 27 - 28 May 2008
[2]Dietrich, R.-U., Oelze, J., Lindermeir, A., Anode side chemical reaction schema and electrochemical performance evaluated on a commercial SOFC stack, Fuel Cells Science & Technology 2008, Scientific Advances in Fuel Cell Systems, Copenhagen, Denmark, 8 - 9 October 2008
[3]Lindermeir, A., Dietrich, R.-U., Oelze, J., Spitta, C., Schönbrod, B., Steffen, M., Evaluation of anode-offgas recycling for a propane operated SOFC-system, H_2-Expo - Internationale Konferenz und Fachmesse für Wasserstoff- und Brennstoffzellen-Technologien, Hamburg, 22 - 23 October 2008
[4]H.C. Starck Ceramics GmbH, Solid Oxide Fuel Cell Products ESC 4, online product brochure, Selb, Germany, 2010
[5]J. Fernández Puga, S. Fleck, M. Mayer, F. Ober, T. Stengel, F. Ebert CFD-Simulation der Strömung in und nach einer Laval-Düse. Chemie Ingenieur Technik (74) 8/2002 S. 1100-1105, WILEY-VCH Verlag GmbH & Co. KGaA, Weinheim, 2002
[6]Somsak Watanawanavet, Optimization of a High Efficiency Jet Ejector by Computational Fluid Dynamics Software. Thesis. Texas A&M University, May 2005
[7]Yinhai Zhu, Wenjian Cai, Changyun Wen, Yanzhong Li, Fuel ejector design and simulation model for anodic recirculation SOFC system. Journal of Power Sources 173 (2007) 437–449, Singapore, Xi'an, China, 19 August 2007
[8]Yinhai Zhu, Wenjian Cai, Yanzhong Li, Changyun Wen, Anode gas recirculation behavior of a fuel ejector in hybrid solid oxide fuel cell systems: Performance evaluation in three operational modes. Journal of Power Sources 185 (2008) 1122–1130, Xi'an, China, Singapore, 23 July 2008

Author Index